Soft Computing Applications

Advances in Soft Computing

Editor-in-chief
Prof. Janusz Kacprzyk
Systems Research Institute
Polish Academy of Sciences
ul. Newelska 6
01-447 Warsaw, Poland
E-mail: kacprzyk@ibspan.waw.pl
http://www.springer.de/cgi-bin/search-bock.pl?series=4240

Esko Turunen
Mathematics Behind Fuzzy Logic
1999. ISBN 3-7908-1221-8

Robert Fullér
Introduction to Neuro-Fuzzy Systems
2000. ISBN 3-7908-1256-0

Robert John and Ralph Birkenhead (Eds.)
Soft Computing Techniques and Applications
2000. ISBN 3-7908-1257-9

Mieczysław A. Kłopotek, Maciej Michalewicz
and Sławomir T. Wierzchoń (Eds.)
Intelligent Information Systems
2000. ISBN 3-7908-1309-5

Peter Sinčák, Ján Vaščák, Vladimír Kvasnička
and Radko Mesiar (Eds.)
The State of the Art in Computational Intelligence
2000. ISBN 3-7908-1322-2

Bernd Reusch and Karl-Heinz Temme (Eds.)
Computational Intelligence in Theory and Practice
2001. ISBN 3-7908-1357-5

Robert John and Ralph Birkenhead (Eds.)
Developments in Soft Computing
2001. ISBN 3-7908-1361-3

Mieczysław A. Kłopotek, Maciej Michalewicz
and Sławomir T. Wierzchoń (Eds.)
Intelligent Information Systems 2001
2001. ISBN 3-7908-1407-5

Antonio Di Nola and Giangiacomo Gerla (Eds.)
Lectures on Soft Computing and Fuzzy Logic
2001. ISBN 3-7908-1396-6

Tadeusz Trzaskalik and Jerzy Michnik (Eds.)
Multiple Objective and Goal Programming
2002. ISBN 3-7908-1409-1

James J. Buckley and Esfandiar Eslami
An Introduction to Fuzzy Logic and Fuzzy Sets
2002. ISBN 3-7908-1447-4

Ajith Abraham and Mario Köppen (Eds.)
Hybrid Information Systems
2002. ISBN 3-7908-1480-6

Lech Polkowski
Rough Sets
2002. ISBN 3-7908-1510-1

Mieczysław A. Kłopotek, Sławomir T. Wierzchoń
and Maciej Michalewicz (Eds.)
Intelligent Information Systems 2002
2002. ISBN 3-7908-1509-8

Andrea Bonarini
Francesco Masulli
Gabriella Pasi
Editors

Soft Computing
Applications

With 147 Figures
and 64 Tables

Physica-Verlag

A Springer-Verlag Company

Prof. Andrea Bonarini
Department of Electronics and Information
Politecnico di Milano
Piazza Leonardo da Vinci 32
20133 Milano
Italy
bonarini@polimi.it

Prof. Francesco Masulli
Department of Computer Science
Genoa University
Via Dodecaneso 35
16146 Genova
Italy

Dr. Gabriella Pasi
National Research Council
Institute for Multimedia Technologies
Via Ampère 56
20131 Milano
Italy
gabriella.pasi@itim.mi.cnr.it

ISSN 1615-3871
ISBN 3-7908-1544-6 Physica-Verlag Heidelberg New York

Cataloging-in-Publication Data applied for
A catalog record for this book is available from the Library of Congress.
Bibliographic information published by Die Deutsche Bibliothek
Die Deutsche Bibliothek lists this publication in the Deutsche Nationalbibliografie; detailed bibliographic
data is available in the Internet at <http://dnb.ddb.de>.

Physica-Verlag Heidelberg New York
a member of BertelsmannSpringer Science+Business Media GmbH

http://www.springer.de

© Physica-Verlag Heidelberg 2003
Printed in Germany

Softcover Design: Erich Kirchner, Heidelberg

SPIN 10894427   88/3130-5 4 3 2 1 0 – Printed on acid-free paper

PREFACE

This volume is a collection of some selected and edited contributions presented at the Fourth Italian Workshop on Fuzzy Logic, held in Milan on September 2001. Although the name of the workshop refers to Fuzzy Logic only (this choice was made to give a continuity to an initiative founded six years ago), the call for papers asked for papers in the more general area of "soft computing" techniques, thus including contributions also concerning the theory and applications of Neural Networks and Evolutionary Computation.

This series of workshops was conceived to the aim of offering to both research institutions and industry a forum where to discuss and exchange ideas concerning the problem of defining "flexible" and "intelligent" systems, able to learn models: data models, the users' behaviour or the environment features in general. Soft Computing techniques are naturally suited and exploited to implement such systems.

The topics covered by the selected papers witness the actual research trend towards an integration of distinct formal techniques for defining flexible systems. The contributions in this volume mainly concern the definition of systems in several application fields, such as image processing, control, asset allocation, medicine, time series forecasting, qualitative modeling, support to design, reliability analysis, diagnosis, filtering, data analysis, land mines detection and so forth.

September 2002

The editors
Andrea Bonarini, Milan
Francesco Masulli, Genoa
Gabriella Pasi, Milan

Contents

Learning Fuzzy Classifiers with Evolutionary Algorithms

Mauro L. Beretta[1] and Andrea G. B. Tettamanzi[2]

[1]Genetica S.r.l.- Via S. Dionigi 15, I-20139 Milano, Italy
E-mail:beretta@genetica-soft.com
[2]Università degli Studi di Milano,Dipartimento di Tecnologie dell'Informazione
Via Bramante 65, I-26013 Crema (CR), Italy E-mail: andrea.tettamanzi@unimi.it

Abstract. This paper illustrates an evolutionary algorithm, which learns classifiers, represented as sets of fuzzy rules, from a data set containing past experimental observations of a phenomenon. The approach is applied to a benchmark dataset made available by the machine learning community.

1   Introduction

Classification, or pattern recognition, is the task of assigning an instance to one of two or more pre-defined classes, or patterns based on its characteristics. A classifier is an object, which computes a mapping from instance data to their (hopefully correct) class. An example of classification is, given a printed character (the instance), to associate it with the letter of the alphabet (the pre-defined class, or the pattern) it represents.

Known patterns are usually represented as class structures, where each class structure is described by a number of features. These features define a feature space, wherein data are defined.

In this paper we illustrate an approach to automatically learning the fuzzy decision rules of a classifier by evolutionary algorithms.

2   The Problem

Classification can be formulated as an optimization task by defining an objective function of the form

$$Z=F(\text{classifier; data set}) \tag{1}$$

which measures the accuracy of a classifier when applied to the data set.

Let us assume that a data set consists of N records, each one with a class attribute $c_i \in \{0, M - 1\}$ where M is the number of classes. Let c^* indicate the classifier's classification of a record. Then, the weighted mean squared error can be calculated as follows:

$$z = \frac{1}{N} \sum_{i=1}^{N} \left[c_i + k(M - 1 - c_i) \right]\!\left(c^* - c_i \right)^2 \tag{2}$$

where parameter k represents a penalty (if $k>1$) or a premium (if $k<1$) for false positives in binary classification tasks. In non-binary classification tasks, k may be set to one (neutral).

3   Fuzzy Rule-based Classifiers

This task can be (and has been) approached in many ways and using many techniques, including decision trees, neural networks, linear classifiers and non-linear statistical classifiers, among the most popular ones. All techniques have their advantages and drawbacks. However, one thing that is often required in practical applications is the interpretability, or explanatory power, of a classifier.

This is the main motivation for evolving classifiers made of decision rules. Moreover, because reality is not sharp and clean and in general decisions taken using crisp thresholds are dangerous in a number of real-world applications, we have moved our attention to fuzzy rules, for their intrinsic interpolative behavior.

The idea of using evolutionary algorithms to design fuzzy systems date from the beginning of the Nineties [5,12] and a fair body of work has been carried out throughout the past decade [6,7,4]. The approach we followed is largely based on the development of a previous work on the evolution of fuzzy controllers [11].

Based on that work, we define a classifier as a rule base, of up to 256 rules, each comprising up to four antecedent and one consequent clause. Up to 256 input variables and one output variable can be handled, described by up to 16 distinct membership functions each. Membership functions for input variables are trapezoidal, while membership functions for the output variables are triangular.

4 The Evolutionary Algorithm

Our approach uses an evolutionary algorithm to evolve fuzzy classifiers of the data set. Evolutionary algorithms are a broad class of optimization methods inspired by Biology that build on the key concept of Darwinian evolution. It is assumed that the reader is already familiar with the main concepts and issues relevant to evolutionary algorithms; good reference books are [3,8,2].

4.1 The Algorithm

An island-based distributed evolutionary algorithm is used to evolve classifiers. An island-based algorithm maintains several populations (islands) wich evolve separately exchanging individuals from time to time. The sequential algorithm executed on every island is a standard generational replacement, elitist evolutionary algorithm. Crossover and mutation are never applied to the best individual in the population.

Encoding

Classifiers are encoded in three main blocks: a set of membership functions for the input variables, a set of symmetric membership functions, represented by (area, center of mass) pairs for the output (or classification) variable, and a set of rules.

A single input variable membership function is defined by four fixed-point numbers, each fitting into a byte. Output variable membership functions are assumed to be symmetrical, and thus can be described using just two parameters: their area and center of mass. A rule is a list of up to four conjoint antecedent clauses (the IF part) and a consequent clause (the THEN part). A clause is represented by a pair of indices referring respectively to a variable and to one of its linguistic values, i.e., a membership function.

Initialization

The population can be seeded either with hand-written or otherwise already existing classifiers or with new random ones. A new random individual is created according to the following algorithm:

1. Each input variable has at least one linguistic value; the number of additional values is determined by sampling from a truncated exponential distribution with mean three;

2. The shapes of the membership functions for the input variables are determined by randomly extracting a center C with uniform probability over the variable definition interval, and a spread σ from an exponential distribution with mean $1/4$ of the variable range. The four numbers defining the trapezoid, a, b, c and d are assigned as follows:

$$a = C - 2/3\sigma,$$

$$b = C - 1/3\sigma,$$

$$c = C + 1/3\sigma, \tag{3}$$

$$d = C + 2/3\sigma$$

The exponential distribution is used to determine the spread because it is the zero-information probability distribution for a non-negative random variable.

3. At least two output membership functions have to be present for each output variable; the number of additional linguistic values for each output variable is determined by sampling from a truncated exponential distribution with mean three.

4. The centers of mass for the output variable are randomly extracted in the $[0, M-1]$ interval; the areas are extracted at random such that they correspond to a triangular membership function whose base is entirely contained in that range;

5. At least M rules have to be present, using at least M different output linguistic values; the number of additional rules in the rule base is determined by sampling from a truncated exponential distribution with mean six;

6. The rules are generated according to the following algorithm:

(a) for each input variable, a fair coin is flipped to decide whether to include it in the antecedent part, not exceeding four variables;

(b) for each selected input variable, a linguistic value is extracted among those defined;

(c) an output variable and its linguistic value are extracted for the consequent part of the rule.

Recombination

The recombination operator is designed to preserve the syntactic legality of classifiers. A new classifier is obtained by combining the pieces of two parent classifiers. Each rule of the offspring classifier can be inherited from one of the parent programs with probability *1/2*. When inherited, a rule takes with it to the offspring classifier all the referred linguistic values with their membership functions. Other linguistic values can be inherited from the parents, even if they are not used in the rule set of the child classifier, to increase the size of the offspring so that their size is roughly the average of its parents' sizes.

Mutation

Like recombination, mutation also produces only legal classifiers. Mutation can result in one or more of the following changes, with probability given by the mutation rate, p_m, identical and independent for each component of the genotype:

- a new linguistic value with a random membership function is added to an input variable;
- a linguistic value whose membership function is not used in the rules is removed from an input variable;
- a membership function is perturbed as follows: each of the four points defining the trapezoid can be moved, with probability p_m, to a new random position between the previous and next points;
- a new linguistic value, with random area and center of mass, is added to an output variable;
- a linguistic value whose area and center of mass are not used in the rules is removed from an output variable;
- an area-center of mass pair is perturbed as follows:

1. a standard deviation σ for the perturbation is extracted from an exponential distribution;
2. a new center of mass is extracted from a truncated normal distribution with mean the old center of mass and standard deviation σ;
3. a new area is extracted from a truncated exponential distribution with mean the old area, such that it corresponds to a triangular membership function entirely contained in the range of the relevant output variable;

- a new random rule is added to the rule set; the new rule is generated as follows:
   1. for each input variable, a fair coin is flipped to decide whether to include it in the antecedent part, not exceeding four variables;
   2. for each selected input variable, a linguistic value is extracted among those defined;
   3. an output variable and its linguistic value are extracted for the consequent part of the rule;
- a rule is removed from the rule set;
- a rule gets a random antecedent clause predicating an input variable not yet used added to it;
- an antecedent clause is removed from a rule;
- the predicate of an antecedent clause is modified by randomly extracting one of the linguistic values defined for the relevant input variable;
- the predicate of the consequent clause of a rule is modified by randomly extracting one of the linguistic values defined for the relevant output variable.

Migration

Migration is responsible for the diffusion of genetic material between populations residing on different islands. At each generation, with a small probability (the migration rate), a copy of the best individual of an island is sent to all connected islands and as many of the worst individuals as the number of connected islands are replaced with an equal number of immigrants.

Moreover, an immigrant whose fitness is lower than the lowest fitness in the island population is always discarded.

4.2 Fitness

Instead of having the evolutionary algorithm work with the objective function in Equation 2 directly, after some experiments, we found more effective to formulate the fitness function (to be maximized) as

$$F = kN - \sum_{i=1}^{N} \frac{c_i + k(M-1-c_i)}{M-1} \left| c_i - c^* \right| \tag{4}$$

It can be easily verified that any classifier maximizing the function in Equation 4 will also minimize the objective function in Equation 2. However, Equation 4 has a better capability of distinguishing between classifiers with similar values of the objective function, which, for evolutionary purposes, works better.

4.3 Selection

Elitist linear ranking selection, with an adjustable selection pressure, is responsible for improvements from one generation to the next. Overall, the algorithm is elitist, in the sense that the best individual in the population is always passed on unchanged to the next generation, without undergoing crossover or mutation.

5 Experiments and Results

We tried our algorithm on the Johns Hopkins University Ionosphere data set, part of the UCI Repository Of Machine Learning Databases and Domain Theories [9].

5.1 Data Set

This radar data measure free electrons in the ionosphere. Positive returns are those showing evidence of some type of structure in the ionosphere. Signals were processed using an autocorrelation function whose arguments are the pulse time and its number. There were 17 pulse numbers for the array used. Instances in this dataset are described by 2 attributes (real and imaginary part) per pulse number.

The Ionosphere dataset was investigated using back-propagation and the perceptron training algorithm [10]. Using the first 200 instances for training, which were carefully split in almost 50% positive and 50% negative, it was found that a linear perceptron attained 90.7% accuracy, a non-linear perceptron attained 92%, while back-propagation attained an average of over 96% accuracy on the remaining 150 test instances, consisting of 123 positive and only 24 negative instances. Accuracy on positive instances was reportedly much higher than for negative instances. David Aha briefly investigated this dataset [9] and found that nearest neighbor attains an accuracy of 92.1%, that Ross Quinlan's C4 algorithm attains 94.0% (without windowing), and that IB3 [1] attained 96.7%.

5.2 Experimental protocol

We carried out four tests:
1. training on the first 200 instances, validation on the remaining 151 instances, as in [10];
2. training on the first 234 instances, validation on the remaining 117 instances;
3. training on the first 117 and the last 117 instances, validation on the second 117 instances;
4. raining on the last 234 instances, validation on the first 117 instances.

We were interested in testing the generalization capability of our learning algorithm, as well as its robustness against training datasets where the balance between positive and negative cases is not tuned (which is by far the most common case in real-world applications).

5.3 Results

The number of *perfect* (i.e., zero error) matches attained by the best classifiers produced in a single run of 1,000 generations (a few hours on state-of-the-art PCs) are summarized in Table 1. A classifier may match a record only to a certain extent (i.e., with a small error): while that prediction is still useful, the record's contribution to fitness will get weaker with the square of the error.

Accuracy of the classifier can vary, in this binary classification task, depending on where the cutoff threshold between the two classes is placed. Table 2 shows the number of correct classifications when a cutoff threshold of 0.5 is chosen.

Test		1		2		3		4	
Training	%	163/200	81.5	218/234	93.2	226/234	96.6	215/234	91.9
Validation	%	107/151	70.9	92/117	78.6	94/117	80.3	83/117	70.9

Table 1. Number of instances correctly classified without uncertainty over the total number of instances in the dataset. Note that this is only a lower bound on accuracy.

Test		1		2		3		4	
Training	%	198/200	99.0	232/234	99.1	231/234	98.7	226/234	96.6
Validation	%	150/151	99.3	116/117	99.1	116/117	99.1	95/117	81.2

Table 2. Number of instances correctly classified with a cutoff threshold of 0.5.

A quick inspection of the results in Table 2 reveals a striking generalization capability for Tests 1, 2, and 3. Test 4 proved to be the hardest, as it could have been easily expected. Not surprisingly, Test 1 was the one that gave the best results, since a well balanced dataset was used for training.

Overall, our approach appears to clearly outperform all other learning algorithms, to our knowledge, that were reported to be applied to the same dataset.

6   Conclusions

The experimental results demonstrate that the proposed evolutionary approach is able to find classifiers with a good generalization capability in a reasonable time. They also show generalization robustness in the face of poorly tuned training datasets (cf. the graceful degradation of results in Test 4 as compared to Test 1).

Acknowledgements

We are grateful to the maintainers of the UCI Repository of machine learning [9] for providing us with the dataset we used in our experiments.

10

References

1. D. W. Aha and D. Kibler. Noise-tolerant instance-based learning algorithms. In *Proceedings of the 11th International Joint Conference on Artificial Intelligence (IJCAI-89)*, pages 794–799. Morgan Kaufmann, 1989.
2. T. Back. *Evolutionary algorithms in theory and practice*. Oxford University Press, Oxford, 1996.
3. D. E. Goldberg. *Genetic Algorithms in Search, Optimization & Machine Learning*. Addison-Wesley, Reading, MA, 1989.
4. C. Z. Janikow. A genetic algorithm for learning fuzzy controllers. *Proceedings of the ACM Symposium on Applied Computing*, New York, 1994. ACM Press.
5. C. L. Karr. Genetic algorithms for fuzzy controllers. *AI Expert*, March 1991.
6. M. Lee and H. Takagi. Embedding apriori knowledge into an integrated fuzzy system design method based on genetic algorithms. *Proceedings of the 5th IFSA World Congress (IFSA'93)*, pages Vol.\ II, 1293–1296, July 4–9 1993.
7. M. Lee and H. Takagi. Integrating design stages of fuzzy systems using genetic algorithms. *Proceedings of the 2nd International Conference on Fuzzy Systems (FUZZ-IEEE'93)*, pages Vol.\ I, 612–617, 1993.
8. Z. Michalewicz. Genetic Algorithms + Data Structures = Evolution Programs. Springer-Verlag, Berlin, 1992.
9. P. M. Murphy and D. W. Aha. The {UCI} repository of machine learning databases and domain theories.
URL: http://www.ics.uci.edu/~mlearn/MLRepository.html, December 1995.
10. V. G. Sigillito, S.P. Wing, L. V. Hutton, and K. B. Baker. Classification of radar returns from the ionosphere using neural networks. *Johns Hopkins APL Technical Digest*, 10:262–266, 1989.
11. A. Tettamanzi. An evolutionary algorithm for fuzzy controller synthesis and optimization. In *IEEE International Conference on Systems, Man and Cybernetics*, volume 5/5, pages 4021–4026. IEEE Systems, Man and Cybernetics Society, 1995.
12. P. Thrift. Fuzzy logic synthesis with genetic algorithms. In R. K. Belew and L. B. Booker, editors, *Proceedings of the Fourth International Conference on Genetic Algorithms*, San Mateo, CA, 1991. Morgan Kaufmann.

Evidence of Chaotic Attractors in Cortical Fast OscillationsTested by an Artificial Neural Network

Rita Pizzi, Marco de Curtis, Clayton Dickson

IRCCS Istituto Nazionale Neurologico C. Besta , via Celoria 11, 20133 Milano, Italy
rita.pizzi@rcm.inet.it

Abstract. A novel ANN architecture, called ITSOM, has been used as a non-linear analysis tool in the study of the cortical fast oscillatory activity that has been correlated to perceptual binding and cellular plasticity.

Simultaneous multirecordings of fast oscillatory activity induced by carbachol in the entorhinal cortex of the guinea pig brain *in vitro* have been processed with ITSOM and compared with standard non-linear analysis tools: correlation dimension, Hurst parameter and recurrence quantification analysis.

Evidence of chaotic attractors in signals after pharmacological stimulus has been shown, indicating self-organization in fast oscillatory activity recorded at distant sites in the entorhinal cortex. The data suggest the existence of functional binding elements in this region, proposed to underlie higher brain functions such as memory and learning.

1 Introduction

Many neurophysiologists have proposed that unity of perception could be connected to the self-organization of the gamma waves (~40 Hz) emitted by the cortical neurons, that in many experiments, in presence of sensorial stimuli, synchronize at distant sites and could therefore induce a functional binding. Fast oscillatory activity in limbic cortices has been proposed to subtend a form of functional binding during higher brain functions, such as memory and learning. Global coherent patterns of neuronal activity are considered the main correlate for conscious experience [1][2][3][4].

Simultaneous microelectrode recordings from distant recording sites in the cortex have led to the possibility of analyzing the mechanism by which the activity of a collection of neurons might be coordinated into a unique pattern.

It has been proposed that sensory cortical neurons interact diffusely and that stimulus-evoked action potentials lead to the emergence of a self-organized pattern of activity as a cortical response to the stimulus [5][6].

It has been also proved that gamma activity in the enthorinal cortex (ERC; [7]) can be reproduced by application of carbachol (50-100 □M) in the guinea-pig brain .

In our experiment the gamma activity has been recorded simultaneously in up to 5 sites separated by about 1 mm in the ERC. Several files have been recorded before, during and after pharmacological stimulation with carbachol (Fig. 1).

In order to evaluate the possible correlations present in data, an unsupervised neural network, called ITSOM (Inductive Tracing Self-Organizing Map) has been chosen because it can show the presence of limit cycles or chaotic attractors in data.

Non-linear analysis tools have been used to quantitatively evaluate the attractors generated by the network , i.e. correlation dimension, Hurst parameter and recurrence quantification analysis[9][10][11].

The same parameters have been used to analyse the original temporal series, and results coming from both kinds of analysis have been compared.

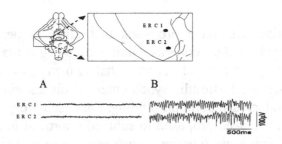

Fig. 1. Schematic drawing of the position of the recording electrodes in the ERC of the guinea-pig brain. In A and B the activity recorded before and after induction of gamma oscillations by carbachol (50 μM) are shown.

2 Methods

The electrophysiological traces were obtained during experiments performed in the isolated guinea-pig brain preparation [8]. Self-organizing neural networks have shown extremely good performances in detecting structures in data. In particular, the Kohonen's SOM is particularly suitable in cases where classification classes are unknown because it acts as a clustering procedure [12].

However, the SOM's performances in case of strictly non-linear and time-variant input are often poor because when the input topology is too tangled, the competitive layer is not able to unfold itself enough to simulate it.

Another problem of the SOM, typical of any clustering algorithm, is the lack of output explication. Once obtained a classification, the user must analyse it, comparing it to the input values in order to extrapolate a significant output. This means that the SOM could be used only by adding an extra module for the output explication.

2.1 ITSOM

It has been observed that if a SOM network is applied to structured data, the temporal sequence of the winning weights tends to repeat itself .

A deeper analysis has shown that such a sequence constitutes cyclic attractors [13] that repeat themselves through the epochs, and that univocally characterize the input element that has determined them.

In this way, due to the countless variety of possible combinations among winning neurons, the configurations allow to finely settle the correct value, even in the case of tangled input topologies, despite of the small number of competitive neurons and their linear topology.

Thus, we developed a network called ITSOM (Inductive Tracing Self-Organizing Network) that takes into account the temporal series of the SOM's winning neurons [14][15][16].

The ITSOM's crucial feature is that this network does not need to be brought to convergence, because if a cyclic configuration is detected, it stabilizes its structure within a small number of epochs, then keeps it steady through time.

After interrupting the network processing phase, an algorithm is needed to codify the obtained cyclic configurations. The algorithm that has shown the best performances is the following, z-score based. The cumulative scores related to each input have been normalized following the distribution of the standardized variable z given by

$$z = (x - \mu)/\sigma$$

where m is the average of the scores on all the competitive layer weights and σ is the mean squared deviation. Once fixed a threshold $0 < \tau \leq 1$, we have put

$$z = 1 \quad \text{for} \quad z > \tau$$
$$z = 0 \quad \text{for} \quad z \leq \tau$$

In this way every winning configuration is represented by a binary number with as many 1 and 0 as many the competitive layer weights.

Then the task of comparing these binary numbers is straightforward. The z-score method has shown steady performances and is computationally not expensive, being linear in the number of the competitive layer neurons.

This allows the network to complete its work within an insensible time, asserting the feasibility of real-time processing. Moreover, this method allows the network to perform a real induction process, because after a many-to-few vector quantization from the input to the weight layer (to be precise, to the cyclic configurations of winning weights), a few-to-many procedure can be performed from the cyclic configurations corresponding to a known reference set to the whole input set (Fig. 2).

In our case the induction process can recognize typical cyclic attractors every time they show themselves in the temporal series.

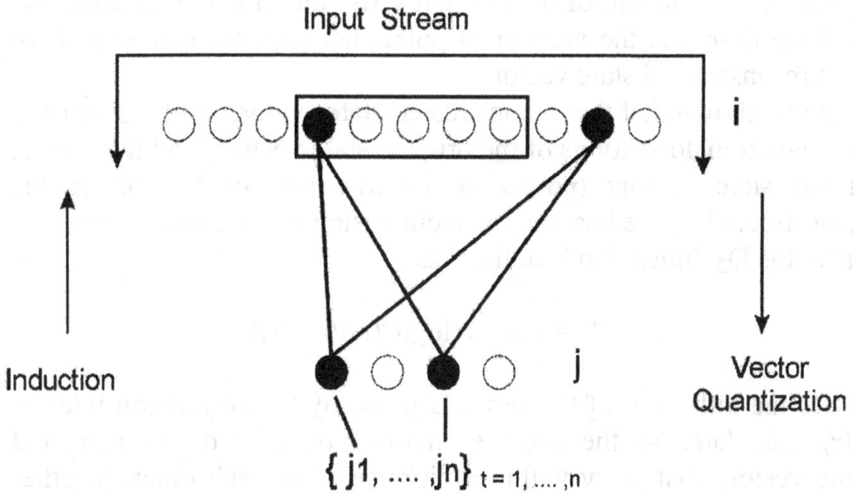

Input Stream

Fig.2. The ITSOM network architecture

Once obtained the temporal series from the neural network, they have be evaluated by means of classical non linear parameters: correlation dimension, Hurst parameter and Recurrence Quantification Analysis.

Correlation Dimension

Grassberger and Procaccia developed a method that allows to settle the so-called correlation dimension D_2 of a time series (that corresponds to a lower bound for the classical Hausdorff fractal dimension) .

Let us apply the so-called delay-time embedding procedure, by substituting in the series each observation of the original signal X(t) with the vector

$$y(i) = x(i), x(i+d), x(i+2d),..., x(i+(m-1)d) ,$$

obtaining as a result a series of vectors of m coordinates in a m-dimensional space:

$$\underline{y} = y(1),y(2),...,y(N-(m-1)d))$$

where N is the length of the original series and d is the so-called lag or delay time, i.e. the number of points between the components of each reconstructed state vector.

It can be shown that the reconstructed state vectors are topologically invariant transformations of the original state vectors, and that the set of the state vectors (points in the m-dimensional state) in the reconstructed space has the same dimension as the system attractor. Then, the D_2 dimension is defined as

$$D_2 = \lim_{r->0} (\log(C(r))/\log(r))$$

where the bulk size of the series is given by the correlation integral $C(r)$, calculated as the average number of pairs of reconstructed state vectors that stay within a distance r of each other. In other words $C(r)$ calculates the mean number of points that stay on the reconstructed attractor within a distance r of each other. If the attractor is a fractal, in a given range of r the logarithm of this average will have a linear relation with the $\log(r)$.

In this region the slope of the curve gives the correlation dimension D_2, that is a measure of the complexity of the system attractor , whereas the correlation integral measures the size of the attractor .

2.2 Hurst Parameter

In a time series the concept of self-similarity is used in a distributional meaning: when viewed at different scales, the distribution of the object remains the same.

A self-similar time series has the property that when aggregated in a shorter series (where each point is the sum of multiple original points) it maintains the same autocorrelation function

$$r (k) = \int x(t) \, x(t+k) \, dt$$

both in the series $X=(X_t: t=0,1,2,...)$ and in the contracted series $X^{(m)} = (X_k^{(m)} : k=1,2,3,...)$, aggregated in blocks of size m. As a result, the self-similar processes show long-range dependence, that is have an autocorrelation function

$$r(k) \sim k^{-\beta} \text{ per } k \rightarrow \infty \quad 0<\beta<1$$

i.e. the function decays hyperbolically.

The degree of self-similarity of a series is expressed using a single parameter., that expresses the speed of decay of the series autocorrelation function and is called the Hurst parameter H :

$$H = 1 - \beta/2.$$

Thus for a self-similar series $\frac{1}{2} < H < 1$.

If $H \rightarrow 1$ the degree of self-similarity increases., therefore the series is self-similar when H is significantly different from $\frac{1}{2}$.

It can be shown that the Hurst parameter is bound to the Hausdorff fractal dimension D (the correlation dimension C(r) is the lower bound of D) by the expression

$$D = 2 - H.$$

2.3 Recurrence Plots

Recurrence Quantification Analysis (RQA) is a new quantitative tool that can be applied to recurrent plots for the analysis of time series reconstructed with delay-time embedding. RQA is independent of data set size, data stationarity, and assumptions on statistical distributions of data.

RQA gives a local view of the series behaviour, because it analyses distances of pairs of points, not a distribution of distances. Therefore unlike autocorrelation RQA is able to analyse fast transients and to localize in time the features of a dynamical variation: for this reasons RQA is ideally suited for physiological systems.

The Taken's theorem states a mathematical relation between the embedding dimension and the real dimension of the attractor of the corresponding dynamical system :

$$n = 2d + 1$$

where n is the best embedding dimension.

Many quantifiers for the RQA evaluation exist; the most significative is DET (determinism), i.e. the percentage of recurrent points that appear in sequence forming diagonal lines in the matrix. DET gives the measure of the space portions in which the system stays for a time longer than expected by chance alone.

The observation of recurrent points consecutive in time (forming lines parallel to the main diagonal) is an important signature of deterministic structure. In fact the length of the longest diagonal (recurrent) line accurately correspond to the value of the maximum Lyapounov exponent of the series. The Lyapounov exponent is another measure of chaoticity , quantifying the mean rate of divergence of neighbouring trajectories along various directions in the phase space of an embedded time series. Time series of chaotic systems have a positive maximum Lyapounov exponent.

3 Results and Conclusions

The signals have been considered simultaneously on all recording sites in order to evaluate the correlation of the activities . The same records have been fed as input of the ITSOM network. The temporal series of winning neurons have been recorded and processed by MATLAB/SIMULINK.

The simulation tool allows to evaluate the possible presence of limit cycles or chaotic attractors, displaying their path in the phase space.

In control conditions (before the activation of fast oscillatory activity; Fig. 1A), the graphics show some organization on the single record, but random or poorly organized patterns in the case of signals coming from more sites.

However, after the induction of fast oscillatory activity with carbachol application, more chaotic patterns are presents, with similar but never identical values, and strongly symmetric shapes. Fig. 3 shows a path in the phase space before and after application of carbachol application.

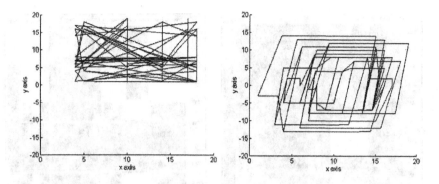

Fig. 3. ITSOM behaviour in the phase space before and after carbachol application, i.e., before and after induction of fast oscillations

In order to quantitatively evaluate the network attractors the Hurst parameter, correlation dimension and recurrence quantitative analysis have been used.

The Hurst parameter, steadily under the value 0.4 before carbachol, sharply grows after the application of carbachol and overcomes the 0.5 threshold, often reaching 0.8. This indicates that signals organize during the stimulation and maintain organization for a while after the stimulus.

The correlation dimension does not seem to be a significant parameter because it keeps steady in the range 2.6-3.2 (using 10 as embedding dimension value) before and after the stimulus: this value seems to be characteristic of this type of signals.

On the other side, the determinism measure of the embedded series, evaluated with the RQA method, confirms the same sharp increase after the stimulus shown by the Hurst parameter, jumping up to over 90% and keeping this high value for a while after the stimulus.

Recurrence plots confirm the existence of extremely regular and evocative patterns in correspondence of high values of H. (Fig. 4).

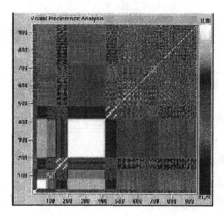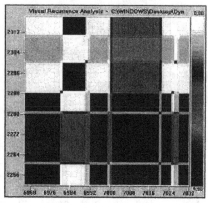

Fig. 4. RQA visualization of two ITSOM attractors

The contextual analysis of the original temporal series with linear methods (coherence computed via power spectrum and cross power spectrum) tested on distant sites shows an increase of the value after the stimulus but keeps lower than 0.4 at any time.

The non-linear analysis on the single record or on two records substantially confirms the ITSOM results (Fig. 5). However in some cases differences exist whose neurophysiological meaning is not known and should be investigated.

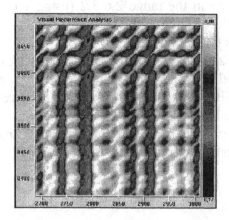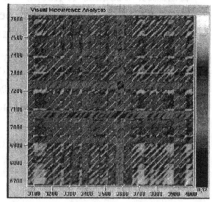

Fig. 5. RQA visualization of two temporal series attractors

More specifically, it must be said that the ITSOM detects self-organized structures more often than the non-linear numerical analysis. This can suggest a better sensitivity of ANNs, though the possibility of false positives should not be excluded. Besides, the ITSOM values of the Hurst parameter result to be often higher than the corresponding ones from the original series. It must be also stressed that, unlike non-linear analysis on the original series, the ANN analysis allowed to simultaneously analyse all the recording sites, testing their possible synchronicity.

Besides, it is also possible, once an organized pattern is detected, to identify its features (represented by its z-score), and identify the same attractor any time the signal or the set of signals generate it.

In conclusion, the present work confirms the existence of a non-linear coherence (in form of chaotic attractors) in the fast oscillations determined by pharmacological manipulation, suggesting that possible functional binding occurs between remote regions in the entorhinal cortex. Our method allows to test the coherence of the whole multisite recording and is suitable for further analysis of possible meanings of the attractors due to the possibility of matching them through time.

References

1. Mitner W.H.R., Braun C., Arnold M., Witte H. and Taub E.: Coherence of Gamma-band EEG Activity as a Basis for Associative Learning, Nature, 397 (1999) 434-436.
2. Rodriguez E., George N., Lachaux J.P., Martinerie J., Renault B. and Varela F.J., Perception's Shadow: Long-distance Synchronization of Human Brain Activity, Nature, 397 (1999) 430-433
3. Varela F.J., Resonant cell assemblies: a new approach to cognitive function and neuronal synchrony, Biol. Res. 28 (1995) 81-95
4. Joliot M., Ribary U. & Llinas R., Human oscillatory brain activity near 40 Hz coexists with cognitive temporal binding, Proc. Natl. Acad. Sci USA 91 (1994) 11748-11751
5. Menon V., Freeman W.J., Spatio-temporal Correlations in Human Gamma Band Electrocorticograms, Electroenc. and Clin. Neurophys. 98 (1996) 89-102.
6. Freeman W.J., Role of Chaotic Dynamics in Neural Plasticity, in: The Self-organizing Brain: from Growth Cones to Functional Networks, Van Pelt J. and Lopes da Silva F.H. (eds) Elsevier (1994)

7. Chrobak J.J., Buzsaki G., Gamma oscillations in the entorhinal cortex of the freely behaving rat. J Neurosci 18 (1998) 388-398.
8. Dickson C.T., Biella G. and de Curtis M., Evidence for spatial modules mediated by temporal synchronization of carbachol-induced gamma rhythm in medial entorhinal cortex, J Neurosci 20 (2000) 7846-7854.
9. Grassberger P., Procaccia I., Measuring the strangeness of a strange attractor, Physica D. 9 (1983) 189-208.
10. Schepers H.E., van Beek J., Bassingthwaighte J.B. Four methods to estimate the fractal dimension from self-affine signals, IEEE Engineering in Medicine and Biology 11 (1992) 57-64.
11. Zbilut JP, Webber CL, Embeddings and delays as derived from quantification of recurrent plots, Phys. Lett. 171 (1992)
12. Kohonen T, Self-Organisation and Association Memory (1990) Springer Verlag
13. Ermentrout B, Complex Dynamics in WTA Neural Networks with slow inhibition, Neural Networks 5 (1992)
14. Pizzi R., Teoria dei Sistemi Dinamici Neurali con Applicazione alle Telecomunicazioni, (Neural Dynamical Systems with Application to Telecommunications), PhD Thesis University of Pavia (1997)
15. Pizzi R., Sicurello F., Varini G., Development of an Inductive Self-organizing Network for the Real-time Segmentation of Diagnostic Images 3th International Conference of Neural Networks and Expert Systems in Medicine and Healthcare, Pisa (1998)
16. Favalli L., Pizzi R., Mecocci A., Non linear Mobile Radio Channel Estimation Using Neural Networks, Proc. of DSP97 Int. Conf. on Digital Signal Processing, Crete (1997).

A Possibilistic Framework for Asset Allocation

Célia da Costa Pereira[1], and Andrea G. B. Tettamanzi[2]

[1] Genetica S.r.l.
Via S. Dionigi 15, I-20139 Milano, Italy
E-mail: dacosta@genetica-soft.com
[2] Università degli Studi di Milano
Dipartimento di Tecnologie dell'Informazione
Via Bramante 65, I-26013 Crema (CR), Italy
E-mail: andrea.tettaman@uni1mi.it

Abstract. This paper is a first step in laying down a theoretical foundation for asset allocation under uncertainty, based on possibility theory. First, fuzzy financial knowledge representation is treated. Then, some issues relevant to financial decision making are approached from the standpoint of possibility theory.

1 Introduction

Asset allocation has become a very complicated problem, as regulatory constraints, individual investor's requirements, non-trivial indices of risk and subjective quality measures are taken into account, together with multiple investment horizons, cash-flow planning and forecasting techniques based on refined econometric models.

This problem has been approached in a probabilistic framework by using sampling techniques to build a tree of possible scenarios for the future from a set of predictions (in the form of joint probability distributions), and an evolutionary algorithm was used to optimize an investment plan against the desired criteria and the possible scenarios [2].

Even though such an approach gave interesting results, we felt that an approach based on probabilistic predictions only is not fully satisfactory, for reasons illustrated below, and that a framework grounded on possibility theory would provide a much more reliable foundation for asset allocation. An approach using fuzzy logic for portfolio selection has be proposed by Ramaswamy in [11]. This work is structured in two parts: Section 2 discusses how uncertainty in financial knowledge can be modeled using fuzzy logic; Section 3 deals with how to use uncertain knowledge to evaluate decisions. Section 4 suggests future directions of research.

2 Uncertainty Representation in Financial Knowledge

2.1 Vagueness, Uncertainty and Imprecision in Finance

Knowledge in Finance is constituted essentially of the views the experts have on the behavior of the market. These views like "index A has good prospects for growth"' are described by the economists as being uncertain, vague, or imprecise in terms of both the direction and the size of market moves.

In the most known method of portfolio management due to Markowitz [12], the *risk* of an investment is used as a synonym for *uncertainty*, and it is measured as the *variance* of a (normally distributed) random variable. The intuitive sense of this hypothesis is that high fluctuations mean greater unpredictability and therefore greater risk. This choice might be explained by the fact that main-stream economists are well acquainted with probability theory, and this has been their main (if not their only) tool to work with uncertainty.

To begin with, it has been shown that variance does not capture the essence of risk for an investor [9]. Besides, some economists make a technical distinction between risk, as a situation in which there are well defined, unique, and generally accepted objective probabilities, and uncertainty [7]. We believe that the scope of uncertainty in this sense is much broader in finance than it is generally acknowledged.

We must draw a clear distiction between uncertainty, vagueness, and imprecision. Uncertainty has to do with telling which of a set of possible events will happen; in the case of vagueness and imprecision, a variable cannot be characterized by a single value of the domain: a vague value is a linguistic label with a possibility distribution, while an imprecise value is an interval of the domain.

These considerations point out that uncertainty differs from imprecision and vagueness, even though it may sometimes result from them [4]. For example, if an analyst says: "There are good chances that the quarterly revenues of company X will show a very slight acceleration", we have uncertainty, in that it is not certain that the event "the quarterly revenues of company X will show a slight acceleration" will happen, but this event is assigned a "probability" through the phrase "there are good chances"; the event itself vague.

States and Scenarios. We may define a *state s* as an assignment of given values to all the variables describing the system of interest, in our case a market. We will denote by S the set of all possible states of the market. We suppose that there exists a state in S that is completely possible, i.e., whose possibility is one. In the case of total ignorance, all states are completely possible.

Although the concept of state is fully appropriate to describe market conditions in the past or in the present, which are known exactly, when reasoning about the future it is convenient to deal with more macroscopic entities, like scenarios.

The word *scenario* is widely used by economists and market analysts to describe their predictions or hypotheses about market behavior. Here, we want to assign to this word a precise meaning, namely, a set of conditions that delimit a conceivable set of states of affairs in the future. For example: "inflation will grow and GNP will increase by less than 3%". As we can see, the formulation of a scenario concentrates on a few variables of interest, and it does not say anything about other variables that are not relevant to the point the analyst wants to make.

The case for fuzzy logic. Although economists normally state their arguments in probabilistic terms, in many cases it makes no real sense to speak of probability, for the repeatability of events, which is typical of the frequentist approach, is lacking. The most commonly used theory of decision-making under uncertainty is subjective expected utility theory, which implies that an individual has beliefs over uncertain events, which have the form of probabilities. These are referred to as *subjective probabilities*.

However, it does not seem plausible that people are always able to assign probabilities to relevant events when they are faced with a problem involving uncertainty, especially if this uncertainty stems from vague or imprecise information. As a matter of fact, there is evidence that people do not have subjective probabilities in some situations involving uncertainty [1]. Moreover, this approach, which would appear more plausible at first sight, is dangerous, for those probabilities are employed in exact calculations and may give the (false) impression that the results thus obtained are exact.

Alternative theories have been proposed by economists to explain uncertainty-averse behavior, for instance complete ignorance, convex sets of probabilities, and non-additive probabilities [7, 8]. These theories are anyway grounded on classical probability theory.

A more promising approach is made possible by fuzzy logic [14], which allows us to model not only uncertainty (to which probability could be suited), but also vagueness and imprecision. Fuzzy sets lead to a gradual theory of uncertainty, which differs in its purpose from the one of frequentist probability. This is possibility theory, where gradual notions of possibility and necessity are present [6].

2.2 Modeling

Two basic types of knowledge are of interest for asset allocation:
1. knowledge about the states of the world at any given time;
2. knowledge about the laws that describe either how the state of the world can change in time, or what relations are there among variables, or both.

Knowledge about the States. This type of knowledge involves in the first place the definition of a number of scenarios. The vagueness and imprecision in the definition of a scenario is modeled in the most natural way by using fuzzy sets. In particular, a scenario will be defined by assigning a linguistic value to some of the relevant (*linguistic*) variables, much like *frames* [10] are defined in terms of *slots*.

In the second place, uncertainty about which scenario will happen in the future can be modeled using possibility theory i.e., as a possibility distribution over the scenarios. Probability distributions can also be used whenever it make sense to use them, considering the remarks made in Section 2.1.

A *fuzzy scenario* C is an object

$$C = \begin{bmatrix} x_1 & is & A_1 \\ & \vdots & \\ x_n & is & A_n \end{bmatrix}, \tag{1}$$

where $x_1, \ldots, x_n$ are n linguistic variables defining the scenario, and $A_1, \ldots, A_n$ are the fuzzy sets of possible values for $x_1, \ldots, x_n$ (i.e., *their linguistic values*). C then represents the fuzzy set $A_1 \times A_2 \times \ldots \times A_n \subseteq S$. For convenience of notation, we will not make a difference between a scenario and the fuzzy set of states it represents.

We will refer to variable x_i of scenario C as $C.x_i$; therefore, Equation 1 will be equivalent to the fuzzy proposition

$$C.x_1 \text{ is } A_1 \text{ AND } \ldots \text{ AND } C.x_n \text{ is } A_n. \tag{2}$$

Let $(x_1, \ldots, x_n)$ be a possible state of the market. Then, its degree of membership in scenario C is given by

$$\mu_C(s) = \min_{i \in \{1, \cdots, n\}} \mu_{A_i}(x_i). \tag{3}$$

We will denote by $\Pi(C)$ the possibility of scenario C, by $Pr(C)$ the probability of the same scenario, if applicable, and by $\pi(s)$ the possibility of state $s \in S$. Thus,

$$\pi(s) = \min_{C \subseteq S} \min(\Pi(C), \mu_C(s)), \tag{4}$$

(we are assuming that $\Pi(C)$ is part of the given knowledge) and

$$Pr(C) = \sum_{s \in S} \mu_C(s) Pr(s). \tag{5}$$

Knowledge about the Laws The laws governing financial pheno-
mena can be synchronic (e.g., "all telecommunications stocks tend
to behave more or less in the same way") or diachronic (e.g., "index
Y is mean-reverting").

Both types of laws can be formally modeled as relations between
financial linguistic variables, and can be expressed as fuzzy
relations. A convenient way of representing fuzzy relations is by
means of fuzzy IF-THEN rules, for which there are well-established
inference methods [15].

3 Evaluation of Financial Strategies under Uncertainty

As a pre-requisite for the evaluation, we need to determine, on the
basis of the current state of the world and of the available
knowledge, which are the most plausible scenarios for the next
period. This can be done by generating all conceivable scenarios one
by one and calculating their possibility. This will yield a ranking of
scenarios, of which, for efficiency reasons, only the first m will be
considered. Strategy evaluation will take place against those m most
plausible scenarios. Of course, the greater m, the more accurate will
be our evaluation, but the more time-consuming it will be.

3.1 Investment Strategies

For the purposes of this work, we can equate an investment strategy
with a *portfolio*, i.e, an allocation of an investment among N asset
classes. An *asset class* is a form of investment. For the sake of
generality, we assume that one of the N asset classes is *cash* (which,
in fact, one would rather define as "no investment"). A portfolio
may be represented as a vector w in the standard simplex W,
satisfying the constraints, for all $i = 1, ..., N$,

$$w_i \geq 0, \tag{6}$$

$$\sum_{i=1}^{N} w_i = 1 \cdot \tag{7}$$

We assume a one-period investment with fixed horizon, whereby an amount of money is invested in the current period in a portfolio, and in the next period the portfolio is transformed back into money, by selling all non-cash assets at market price.

3.2 Fuzzy Utility

The outcome of an investment strategy can be evaluated by means of a utility function, which maps the return of the investment at the final period to a real number, according to some investment objective.

For example, suppose that an investor wants to achieve a return of θ or more from the investment; the investor associates a constant marginal utility to increased return; also, the investor is risk-averse and associates a disutility with underperformance which grows with the square of the shortfall. Then, the investor's utility function will be something like

$$U(R(w);\theta) = R(w) - \lambda[R(w) < \theta](\theta - R(w))^2, \tag{8}$$

where $R(w)$ is the return of portfolio w at the investment horizon, and λ is a trade-off coefficient between risk and return. $R(w)$ depends on the vector r of returns of asset classes at the moment they are sold, which in turn depends on the state of the world s (as defined in Section 2.1) in the period in which the investment ends:

$$R(w) = w \cdot (1 + r(s)). \tag{9}$$

Therefore, utility depends on the investment strategy chosen w, on the state of the market s at the time the investment ends, and is parameterized by any objective θ:

$$U(R(w);\theta) \equiv U(w,s;\theta). \tag{10}$$

Now, in general, the knowledge possessed at the time an investment strategy is to be chosen will result in a possibility distribution over the states $s \in S$. Therefore, instead of knowing in advance the exact utility of every strategy, we will have to be satisfied with a possibility distribution over the utilities, which we can conveniently define *fuzzy utility*. By applying the extension principle, we can write

$$\Pi(u) = \sup_{s:U(w,s;\theta)} \pi(s). \tag{11}$$

Substituting Equation 4 in Equation 11 yields

$$\Pi(u) = \sup_{s:U(w,s;\theta)} \min_{C \subseteq S} \min(\Pi(C), \mu_C(s)). \tag{12}$$

which relates the possibility distribution of utility for a given strategy to the possibilities of the defined scenarios.

3.3 Optimal Investment Strategies

The considerations made in the previous subsection allow us to evaluate the fuzzy utility for any given investment strategy in the light of the knowledge available at the time an investment strategy has to be chosen. Now we want to find the best investment strategies in term of their fuzzy utilities. Basically, we must choose, among all $w \in W$, a portfolio w^*, such that its fuzzy utility $\Pi_{w^*}(u)$ is "better", or "preferable" to the fuzzy utility $\Pi_w(u)$ of any other w. In order to do this, we need to be able to compare possibility distributions over the reals (i.e., fuzzy sets of real numbers) with one another.

Various methods to compare fuzzy numbers have been proposed in the literature [5]. Fuzzy utilities, however, are not always fuzzy numbers. Besides, a comparison method makes sense only with respect to a specific application [3].

In general, we have to admit the possibility that some fuzzy utilities are not comparable, so that we can establish only a partial ordering of fuzzy utilities. Therefore, we have to restate our objective by saying that we must find the set of all non-dominated investment strategies, whose fuzzy utilities are not comparable with one another, but are not "worse" than the fuzzy utility of any other investment strategy.

3.4 Ranking Fuzzy Utilities

Here, we propose three methods to compare fuzzy utilities, which are designed specifically for asset allocation.

Weighted Pessimistic Ranking This method relies on the definition of a relative *merit* of a fuzzy utility U^1 with respect to another fuzzy utility U^2, as a weighted comparison of the infima of all their α-cuts:

$$\tau(U^1 > U^2) = 2 \int_0^1 \alpha [\inf(U_\alpha^1) > \inf(U_\alpha^2)] d\alpha. \tag{13}$$

This merit $\tau(U^1 > U^2)$ may be interpreted as the truth degree of proposition "U^1 is better than U^2". According to this ranking, a fuzzy utility U^1 will be preferable to another fuzzy utility U^2 if its worst utility (weighted by its possibility) is higher than the worst utility compatible with U^2.

Weighted Optimistic Ranking This method is dual to the weighted pessimistic ranking: the relative merit is defined as

$$\tau(U^1 > U^2) = 2 \int_0^1 \alpha [\sup(U_\alpha^1) > \sup(U_\alpha^2)] d\alpha. \tag{14}$$

3.5 Weighted Center of Gravity Ranking

This methods is based on the center of gravity (cfr. the first ranking function proposed by Yager [13]).

$$G = \frac{\int_{-\infty}^{\infty} u\Pi(u)du}{\int_{-\infty}^{\infty} \Pi(u)du}. \qquad (15)$$

The idea is to prefer the fuzzy utility whose center of gravity, weighted by its height (i.e., maximum possibility), is greater:

$$U^1 > U^2 \equiv h(U^1)G(U^1) > h(U^2)G(U^2). \qquad (16)$$

3.6 Combined Method

The three ranking methods proposed above can be combined at least in two ways:
– hierarchically: apply weighted pessimistic ranking first, then, in case of tie, apply weighted center of gravity ranking, then, if there is still a tie, apply weighted optimistic ranking;
– according to a majority rule: apply all three methods in parallel, then take the advice suggested by the majority.

4 Conclusions and Future Work

The work presented above should be considered a starting point for a series of theoretical and practical investigations in the field of asset allocation and finance. A number of extensions are possible (or rather necessary) in order to apply this possibilistic framework to problems of practical relevance:
– we have dealt with the simplest case of asset allocation, namely one-period, fixed-horizon asset allocation; the same ideas could be extended to the case of multi-period, variable-horizon asset allocation;

– we have considered the case in which only possibilities can be assigned to scenarios; an interesting extension would consist n investigating the case in which probabilities are known for some scenarios, and possibilities for other scenarios;
– we have not treated the algorithmic aspects of the framework, i.e., how do we select the most plausible scenarios? How do we use them to find the non-dominated set of investment strategies? How do we do that efficiently?

References

1. P. Anand. Analysis of uncertainty as opposed to risk: An experimental approach. *Agricultural Economics*, 4:145--163, 1990.
2. S. Baglioni, D. Sorbello, C. da Costa Pereira, and A. Tettamanzi. Evolutionary multiperiod asset allocation. In Whitley D, D. Goldberg, E. Cantú-Paz, L. Spector, I. Parmee, and H.-G. Beyer, editors, *Proceedings of the Genetic and Evolutionary Computation Conference GECCO 2000*, pages 597--604, S. Francisco, CA, 2000. Morgan Kaufmann.
3. G. Bortolan and R. Degani. A review of some methods for ranking fuzzy subsets. *Fuzzy Sets and Systems*, 15:1--19, 1985.
4. D. Dubois, W. Ostasiewicz, and H. Prade. Fuzzy sets: History and basic notions. Technical Report 27 R, IRIT, June 1999.
5. D. Dubois and H. Prade. Ranking fuzzy numbers in the setting of possibility theory. *Information Sciences*, 30:183--224, 1983.
6. D. Dubois and H. Prade. *Théorie des possibilités*. Masson, 1988.
7. D. Kelsey. The concept of uncertainty. *Greek Econ, Rev.*, 17(2):61--82, 1995.
8. D. Kelsey and J. Quiggin. Theories of choice under ignorance and uncertainty. *Journal of Economic Surveys*, 6(2):134--153, 1992.
9. A. Loraschi and A. Tettamanzi. An evolutionary algorithm for portfolio selection within a downside risk framework. In C. Dunis, editor, *Forecasting Financial Markets*, Series in Financial Economics and Quantitative Analysis, pages 275--285. John Wiley and Sons, Chichester, 1996.
10. Marvin Minsky. A framework for representing knowledge. In Patrick H. Winston, editor, *The Psychology of Computer Vision*. 1975.
11. S. Ramaswamy. Portfolio selection using fuzzy decision theory. Technical Report 59, Bank For International Settlements, November, 1998.
12. R. Schoenberg. Markowitz portfolio analisis for the individual investor. Technical report, RJS Software Co., Seatle, WA 98121, 1994.
13. R. R. Yager. A procedure for ordering fuzzy subsets of the unit interval. *Information Science*, 24:143--161, 1981.
14. L. A. Zadeh. Fuzzy sets. *Information and Control*, 8:338--353, 1965.
15. L. A. Zadeh. The calculus of fuzzy if-then rules. *AI Expert*, 7(3):22--27, March 1992.

Efficient Low-resolution Character Recognition Using Sub-machine-code Genetic Programming

Giovanni Adorni*, Stefano Cagnoni, Marco Gori^ and Monica Mordonini

Dipartimento di Ingegneria dell'Informazione, Università di Parma,
*Dipartimento di Informatica, Sistemistica e Telematica, Università di Genova
^Dipartimento di Ingegneria dell'Informazione, Università di Siena

Abstract. The paper describes an approach to low-resolution character recognition for real-time applications based on a set of binary classifiers designed by means of Sub-machine-code Genetic Programming (SmcGP). SmcGP is a type of GP that interprets long integers as bit strings to achieve SIMD processing on traditional sequential computers. The method was tested on an extensive set of very low-resolution binary patterns (of size 13 x 8 pixels) that represent digits from 0 to 9. Ten binary classifiers were designed, each corresponding to a pattern class. In case of no response by any of the classifiers, a reference LVQ classifier was used. The paper compares the resulting classifier with a reference classifier, showing an almost 10-fold improvement in speed, at the price of a slightly lower accuracy.

Introduction

Genetic Programming (GP) [1,2] is an Evolutionary Computation (EC) paradigm in which individuals are programs, typically encoded by syntactic trees or, equivalently, prefix-notation functions like LISP functions. GP is usually much more computationally intensive than Genetic Algorithms (GAs), although the two evolutionary paradigms share the same basic algorithm. The higher requirements in terms of computing resources with respect to GAs are essentially due to the much wider search space and to the higher complexity of the decoding process and of the crossover and mutation operators. Therefore, there is great interest in developing new variants of GP that improve the computational efficiency of the paradigm. One such variant, Sub-machine-code GP (SmcGP), has been recently proposed by Poli [3,4].

Sub-machine-code GP is aimed at exploiting the intrinsic parallelism of sequential CPUs. Inside a sequential N-bit CPU (where typically N=32 or N=64), each bit-wise operation on integers is performed by concurrently activating N logical gates of the same kind. Therefore the application of a sequence of bit-wise logical operators to an integer can be seen as a parallel execution of a program on N 1-bit operands in parallel, according to the Single Instruction Multiple Data (SIMD) paradigm.

Such an approach can speed up Genetic Programming (GP) by making it possible to evaluate several fitness cases at the same time, if the teaching input of the fitness cases is binary. When the input patterns of the fitness cases are long bit strings, SmcGP also makes it possible to process an N-bit slice of each input at the same time. In this case, besides speeding up evolution, SmcGP produces programs that are intrinsically parallel and therefore also computationally very efficient.

Because of these properties, SmcGP can be used effectively in all those applications in which the same operations must be performed at one time on a large number of small chunks of data, that can therefore be packed into a long integer variable. This is the case for binary-pattern recognition or binary-image processing, especially when processing of 2D patterns can be performed line-wise or block-wise.

We have experimented SmcGP within a project aimed at developing a vision-based car-plate recognition system (APACHE, Automated PArking CHEck) [5,6]. The system comprises three modules: i) plate detection, in which a region of interest containing a plate is extracted from the input image, ii) character extraction, in which characters and other symbols that compose the plate are isolated and rescaled, and iii) symbol recognition, in which the symbols extracted by the previous module are classified.

To evaluate SmcGP potential in this application domain, such a technique has been used to partially re-design the third module. In particular, we have developed a classifier for digits (2D binary patterns representing the numbers from 0 to 9), comparing it with a modified Learning Vector Quantization (LVQ) classifier [7,8] that is presently used in the APACHE system. This classifier has been used as the reference for performance evaluation.

In the following we describe our approach and report and comment on the results we have achieved on a wide data set of real plate images, in terms of both recognition accuracy and processing speed.

Using SmcGP for license-plate character recognition

At first glance, character classification for license-plate recognition seems to have much in common with Optical Character Recognition (OCR). However, differently from OCR, in which character classification occurs after a text has been scanned at high resolution, license-plate recognition analyses only very low-resolution patterns, possibly altered by optical distortions and perspective effects. Furthermore, in OCR, the image in which characters are detected is by far more structured than the outdoor scene from which characters belonging to a license plate are extracted.

In the application for which we have developed the character classifier described in this paper, characters that are to be subsequently classified are extracted from regions that can be located anywhere within a 512 x 384 pixel image, such as the ones shown in figure 1. Each character is then re-scaled down to a 13 x 8 pixel two-dimensional binary pattern for classification.

The main specifications for automatic license-plate recognition systems can be derived from these considerations. Besides high recognition accuracy, they include robustness with respect to lighting conditions and real-time performance [9].

In this paper we focus on the application of SmcGP to low-resolution character classification, and describe how we designed a set of binary classifiers to perform such a task. As will be pointed out in the conclusions, also other modules of the APACHE system could benefit from the application of SmcGP. We are planning to implement them using such a GP paradigm, since most of the computation load required from APACHE is devoted to processing binary images, that is the case in which SmcGP can produce its best results in terms of computation speed-up.

Fig. 1. Two typical shots acquired by the APACHE system

For the kind of license plates APACHE is presently able to deal with (the last 2 generations of Italian plates [10]), there are rules that specify the syntax of the plate, i.e., where letters or digits are to be found in it. Therefore, the digit classifier can be run independently of the specific letter classifier when any one symbol, whose position within the plate has been correctly identified, has to be recognized. For this reason, as a first step in the evaluation of SmcGP potentials in character classification, only the digit classifier has been re-implemented. The problem that has been tackled is therefore a 10-category classification problem.

Classification strategy

A set of 10 binary classifiers, corresponding to each of the 10 decimal digits, has been developed. The output of each classifier is 1, when the input pattern represents the corresponding digit, or 0 otherwise. The final classification can be derived from the analysis of the response of all classifiers.

The desired behavior of such a system is the one in which only the classifier corresponding to the class of the input pattern outputs 1, while all others give 0 as output. As pointed out in [11], where this kind of classification architecture is described in details with reference to the use of classifiers evolved by GP, there are, quite intuitively, two cases in which this strategy fails. The first is when no classifier produces 1 as output, while the other one is when more than one classifier produce 1 as output.

Having a good classifier already available (the LVQ reference classifier), our choice was to use it to disambiguate such undecidable cases. As described below, the fitness function that was used to evolve the classifier set was chosen such that the number of ambiguously-classified patterns, whose recognition had to be deferred to the LVQ classifier, was as small as possible.

Encoding of input data

To apply SmcGP to character classification on a 32-bit architecture (Pentium III PC running Linux 2.2 with *gcc* 2.95 compiler) each pattern was encoded as four 32-bit integers. Since the number of bits in each pattern (104) is not a multiple of 32, the first nine lines of the pattern were encoded as the 24 least significant bits of the first three integers, while the whole fourth integer was used to encode the last four lines of the pattern, as shown in figure 2.

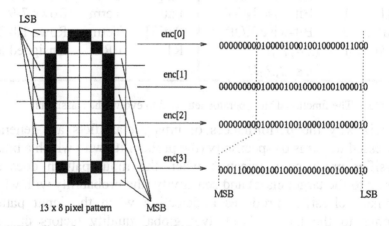

Fig. 2. Encoding of a character as four 32-bit integers

Evolutionary algorithm setup

The classifiers were evolved using *lil-gp1.01* [12], a popular package that implements Koza-like (i.e., LISP-like or tree-like) GP. The function set was composed by the main bit-wise boolean operators and by a set of circular (the LSB is considered to be adjacent to the MSB) shift operators with variable shift direction (left or right) and entity (1, 2, or 4 bits). The terminal set was composed by the 4 unsigned long integers into which the input pattern is encoded, and an unsigned long integer ephemeral random constant (ERC) [1,2], that can take values within the whole range of 32-bit unsigned integers.

To define the fitness function, a major peculiarity of multi-category classification problems must be taken into account. As pointed out in [11], when N mutually-exclusive classes are considered, if samples in the training set are equally distributed among the classes and a binary classifier is used for each class, a training set includes N-1 times as many negative examples as positive ones.

Function set			Terminal Set		
Function	Arity	Notes	Terminal	Type	Notes
AND	2	Bit-wise AND	Pat[0]	Term	Rows 1-3
OR	2	Bit-wise OR	Pat[1]	Term	Rows 4-6
NOT	1	Bit-wise NOT	Pat[2]	Term	Rows 7-9
XOR	2	Bit-wise XOR	Pat[3]	Term	Rows 10-13
SHxn	1	$x \in \{L,R\}$, $n \in \{1,2,4\}$	R1	ERC	Unsigned long

Table 3. The functional and operator sets used to evolve the classifiers

This is why the performances of binary classifiers are generally expressed in terms of specificity (the probability with which a binary classifier produces a rejection when the input pattern does not belong to the target class) and sensitivity (the probability with which a binary classifier produces a detection when the input pattern belongs to the target class), two global quality factors that are independent of the number of examples [13]. When positive and negative examples are equally distributed, the root mean square of sensitivity and specificity is generally a good choice for the fitness function, because it ensures a good balance between the two, often counteracting, factors. In that case, the fitness function is roughly proportional to the number of correct classifications.

However, if the distribution of positive and negative examples is not uniform, such a fitness function allows for N-1 as many false negatives as false positives. Therefore, if maximizing the number of correct classifications is the final goal, the number of incorrect classifications must appear explicitly in the fitness function. After such considerations, the fitness function was defined as

$$F = 1 - \sqrt{ ((FP^2 + FN^2) / (Np + Nn)^2) }$$

where FP is the number of false positives, FN the number of false negatives, Np the number of positive examples in the training set, and Nn the number of the negative examples in the training set. For the above-mentioned property of multi-category classification, this choice leads to very specific classifiers, with very high positive predictivity, i.e., the probability for the classifier of being correct when it outputs 1. This drastically limits the number of ambiguous classifications where more than one classifier produce 1 as output. As reported in section 3, this choice did not boost the number of patterns, which no classifier could detect as belonging to its class, which was also quite low.

Due to the closure requirement of GP, the output of each classifier that operates on unsigned long words, is an unsigned long as well. However, what is needed is a binary output. If one looks upon the unsigned long output as the output of 32 different (even if not independent) binary functions, the output bit can be chosen as the one that maximizes the fitness. Therefore, decoding each individual in the population implies two steps. In the first one, the function encoded by the corresponding tree is decoded. In the second one, each bit of the 32-bit word obtained in the first step is considered as the output of the classifier and fitness is evaluated accordingly. The fitness of each individual is therefore equal to the maximum fitness obtained in the second step.

Evolution was applied to a population of 1000 individuals, with a crossover rate of 80%, a mutation rate of 2% and a reproduction rate of 18%. Tournament selection was used with tournament size equal to 7. Each classifier was evolved for 1000 generations. At least two runs for each classifier were performed. The resulting classifiers were converted from prefix notation, typical of syntactic trees, to infix notation, and translated into C functions to allow for their compilation.

The data set included 11034 patterns collected at highway toll booths, almost uniformly distributed among the ten classes under consideration. The patterns consist of 8-bit grayscale images that have been re-scaled to a size of 13 x 8 pixels, and binarized using a threshold of 128, to allow for their packing into 4 unsigned long words. The training set included 6024 such patterns while the test set included 5010 of them.

Results

The performance of the classifiers was evaluated both in terms of accuracy and computational efficiency. Table 2 reports the response of each classifier in terms of a confusion-like matrix. Each column of the matrix represents the frequency with which the corresponding classifier has produced 1 as output in the presence of patterns of each of the 10 classes; the last row of the matrix reports the number of patterns belonging to each class for each of the two data sets. As can be observed, very few misclassifications (false positives) occurred, while false negatives were slightly more frequent.

	Test Set										Training Set									
	0	1	2	3	4	5	6	7	8	9	0	1	2	3	4	5	6	7	8	9
0	481	2	0	4	2	1	3	0	1	0	551	0	0	0	0	0	0	0	0	0
1	0	491	2	2	1	1	0	3	1	0	0	521	0	1	0	0	0	0	0	0
2	0	1	477	1	7	0	0	0	1	1	0	0	562	0	1	0	0	0	0	0
3	3	2	2	483	1	6	0	0	0	0	0	0	0	735	0	0	0	0	0	0
4	0	8	4	1	480	3	2	0	2	0	0	0	0	0	590	0	0	0	0	0
5	2	1	0	6	0	482	9	0	0	9	0	0	0	0	0	616	1	0	0	0
6	0	0	0	0	3	2	478	0	5	0	0	0	0	0	0	0	596	0	0	0
7	0	8	3	1	2	0	2	475	1	0	0	0	1	0	0	0	0	534	0	0
8	1	0	0	0	0	3	4	0	460	2	0	0	0	0	0	0	0	0	592	0
9	0	1	0	1	1	9	0	3	10	482	0	0	0	0	0	0	0	0	0	673
N	501	501	501	501	501	501	501	501	501	501	556	525	566	739	601	619	599	534	611	674

Table 4. Classification results and distribution of the patterns (N is the number of patterns belonging to the class of the classifier under consideration)

Table 3 reports the specificity and sensitivity of the ten classifiers, along with data related with their computational efficiency. In particular, the table reports the number of nodes in the tree-like representation produced by *lil-gp*, and the processing time in microseconds required by each classifier on a Pentium III-600Mhz PC.

As can be noticed, specificity was always very high (above 99.4%) on both the training and test sets. Sensitivity was also quite good even if generalization was not excellent in some cases, and led to some remarkable differences between the training and the test sets. Processing time varied from 1.2 to 3.99 and was not exactly proportional to the classifier size, since basic bit-wise boolean functions can be virtually run in one clock tick, while circular shift operators are at least three times as demanding, since they are composed by two basic shift and one OR operations.

	0	1	2	3	4	5	6	7	8	9
Specificity (training set)	100.0	100.0	99.98	99.98	99.98	100.0	99.98	100.0	100.0	100.0
Sensitivity (training set)	99.10	99.24	99.29	99.46	98.17	99.52	99.50	100.0	96.89	99.85
Specificity (test set)	99.87	99.49	99.76	99.65	99.62	99.45	99.56	99.87	99.53	99.73
Sensitivity (test set)	96.01	98.00	95.21	96.41	95.81	96.21	95.41	94.81	91.82	96.21
Tree nodes	45	115	290	154	71	195	232	96	304	210
Processing time (μs/PIII600)	1.40	1.80	2.59	1.60	1.60	3.79	3.99	1.20	3.79	3.59

Table 5. Performance of the classifiers

As concerns the two-stage classification strategy, the binary SmcGP-based classifier set was able to classify 99.07% of patterns in the training set, of which 99.98% were correctly classified, and 95.31% in the test set, of which 98.68% were correctly classified. This means that only for a very limited number of cases (less than 5%) was it necessary to demand classification to the less efficient LVQ classifier.

The global accuracy was 98.3% correct classifications, versus 98.68% achieved by the LVQ classifier. The processing time was .15 s on a Pentium III 600 MHz for the test set, which is about 10 times as fast as the LVQ reference classifier, that requires 1.37 s for the same task. It should be noticed that both processing times include time needed to read data from a file, which is smaller (4 long integers to be read per pattern) in the case of the SmcGP-based classifiers than in the case of the LVQ classifier (104 chars to be read per pattern). However, in the latter case, the time required to convert the patterns that cannot be classified in the first stage from the 4-long integer representation to the 104-char representation is also included.

Final remarks

Applying SmcGP to two-dimensional binary pattern classification yielded very good results, both in terms of accuracy and efficiency, even relying on a quite naive classification scheme (one *layer* of binary classifiers). In [14] it was proposed that such a scheme be improved by introducing either a new layer of classifiers, to disambiguate those cases when more than one binary classifiers fire in the presence of a certain pattern, or using some kind of tournament-like strategies to produce the final classification. However, the aim of the experiment described in this paper was to evaluate the basic potentials of SmcGP in this kind of application. The results show that, even in this simple configuration, the SmcGP-based classifiers yields only slightly worse performance than the reference LVQ classifier, that in previous tests had outperformed other kinds of classifiers (e.g., classifiers based on multi-layer perceptrons trained with the backpropagation algorithm). However, the computation efficiency of the SmcGP-based classifier is much better. It could be even up to twice as good if the classifier could be run on a 64-bit architecture. The results obtained are inducing us to extend the use of SmcGP from simple pattern recognition to more general image processing problems. In this regard, the structure of the APACHE plate-recognition system offers new opportunities of application. In particular, it should be noticed that most processing in the APACHE system is performed on binary images, not only in the classification stage but also in the plate-detection and in the symbol segmentation stages. Therefore also the algorithms involved in those stages could be re-written and optimised genetically using SmcGP.

Acknowledgements

This project has been partially funded by ASI (Italian Space Agency) under the ``Hybrid Vision System for Long Range

Rovering" grant and by CNR (Italian National Research Council) under the "ART - Azzurra Robot Team" grant.

References

1. J.Koza. Genetic Programming: On the Programming of Computers by Means of Natural Selection. MIT Press, Cambridge, 1992.

2. W.Banzhaf, F.Francone, J.Keller, and P.Nordin. *Genetic Programming: An Introduction*. Morgan Kaufmann, 1998.

3. R.Poli and W.B. Langdon. Sub-machine-code Genetic Programming. In L.Spector, U.M.O'Reilly W.B.Langdon, and P.J.Angeline, editors, *Advances in Genetic Programming 3*, chapter 13, pages 301-323. MIT Press, 1999.

4. R.Poli. Sub-machine-code GP:New results and extensions. In W.B.Langdon R.Poli, P.Nordin and T.Fogarty, editors, *Proceedings of the Second European Workshop on Genetic Programming - EuroGP'99*, number 1598 in Lecture Notes on Computer Science, pages 65--82. Springer Verlag, 1999.

5. G.Adorni, F.Bergenti, S.Cagnoni, and M.Mordonini. License-plate recognition for restricted-access area control systems. In G.L.Foresti, P.Mähönen, and C.S.Regazzoni, editors, *Multimedia Video-Based Surveillance Systems: Requirements, Issues and Solutions*. Kluwer, 2000.

6. G.Adorni, S.Cagnoni, M.Gori, and M.Mordonini. Access control system with neuro-fuzzy supervision. In *Proc. of the Intelligent Transportation Systems Conference (ITSC2001)*, pages 472-477, 2001.

7. T.Kohonen. Self-organization and associative memory (2nd ed.). Springer-Verlag, Berlin, 1988.

8. S.Cagnoni and G.Valli. OSLVQ: a training strategy for optimum-size Learning Vector Quantization classifiers. In *Proc. of the 1st IEEE World Conference on Computational Intelligence: ICNN94*, pages 762-765, June 1994.

9. J.A.G. Nijhuis, M.H. ter Brugge, K.A. Helmolt, J.P.W. Pluim, L.Spaanenburg, R.S. Venema, and M.A. Westenberg. Car license plate recognition wiht neural networks and fuzzy logic. In *Proc. IEEE Int'l Conf. on Neural Networks*, volume 5, pages 2232--2236, 1995.

10. N.Parker, J.Weeks, and R.Wilson, editors. *Registration plates of the world*. Europlate, 3rd edition, 1995.

11. J.K. Kishore, L.M. Patnaik, V.Mani, and V.K. Agrawal. Application of genetic programming for multicategory pattern classification. *IEEE Trans. on Evolutionary Computation*, 4(3):242--258, 2000.

12. D.Zongker and B.Punch. *lil-gp 1.01 user's manual*. Michigan State University, 1996, available via anonymous ftp from ftp://garage.cse.msu.edu/pub/GA/lilgp.

13. J.P. Egan. *Signal Detection Theory and R.O.C. Analysis*. Academic Press, New York, 1975.

14. G.Adorni, F.Bergenti, and S.Cagnoni. A cellular-programming approach to pattern classification. In W.Banzhaf, R.Poli, M.Schoenauer, and T.C. Fogarty, editors, *Proceedings of the First European Workshop on Genetic Programming(EuroGP98)*, number 1391 in Lecture Notes on Computer Science, pages 142-150, Springer-Verlag, 1998.

Accurate Modeling and NN to Reproduce Human Like Motion Trajectories

Pietro Cerveri[1,2] , Giuseppe Andreoni[1,2], Giancarlo Ferrigno[1,2]

[1] Dipartimento di Bioingegneria, Politecnico di Milano, via Golgi 39 - 20133 Milano, Italy {andreoni, cerveri, ferrigno} @biomed.polimi.it

[2] Centro di Bioingegneria, Politecnico di Milano-Fondazione Pro Juventute Don Carlo Gnocchi ONLUS, via Capecelatro, 66 –20148 Milano, Italy

Abstract: Due to increasingly and massive use of simulation for studying man-workplace interaction, virtual design and CAD systems will be more and more concerning with virtual actors able to emulate the real human motion, with high resemblance. In this work we present some results related to the attempt of synthetically generating human like motion trajectories to drive a model of a human arm. The problem of producing human-like trajectories of complex articulated structures has been re-framed as the issue of computing the parameters of a neural network through a large redundant data set, without involving a mathematical model of either direct or inverse kinematics. In particular, reaching and pointing motor tasks have been considered and tested on a human model. The kinematics has been described in terms of joint angles computed from a set of 3D trajectories of markers, located onto the subject and acquired by means of an opto-electronic motion analyzer. Our approach for movement simulation has been based on a multi-layer perceptron able to predict the trajectory of an arm (three sticks model) in term of joint angles by specifying only the starting and the final position (3D coordinates) of the end-effector. Results, in term of residual errors in training and extrapolation properties, show the reliability of the proposed method.

1 Introduction

Study and design of manned structures and environments have to take into account both static and dynamic features of the man, that is the anthropometrical parameters and the motor performance. Typical approaches use either physical mock-ups or real people, thus being time and money wasting and leading to results strongly dependent on the subject. Recent computer-based techniques allow creating virtual environments modeling both structures and human beings [1,2,3,4]. Additionally, building virtual actors, which operate according to human behaviors, requires to tackle the problem of redundant kinematic systems. Direct kinematics of these structures is straightforward whereas the inverse kinematics, being unconstrained, leads to multiple solutions because of degree of freedom (Dof) redundancy. In this framework, controllers, which involve the use of mathematical models based on inverse kinematics have been largely proposed [5,6,7]. However they require a good calibration of the constraints and sophisticated cost function optimization. Flash and Hogan [8] proposed the 'minimum jerk model', where the cost function is the square of the jerk (time derivative of the acceleration) of the hand position integrated over all the movement. Their analysis was based exclusively on the kinematics of the movements, independently by the dynamic of the muscular and skeletal systems. Uno et al. [10], starting from the consideration that, while performing reaching movements, a human subject moves along an almost straight line trajectory with a bell-shaped velocity profile, proposed a dynamics related mathematical model based on the minimum torque-change criterion. The main drawbacks of these approaches are:

- the limitation to planar movements even though the obtained hand trajectories approximation is quite good;
- the inability to reproduce human-like movements with a reliability that makes them easily distinguishable from real movements;
- the limitation of these parametrical model that are not able to interpret the redundancy in the degrees of freedom (Dof) of the human biomechanical model in the natural movement.

A completely different solution is represented by a complete non-parametric estimator-predictor, adopting a neural network (NN) framework, which exhibits ability to cope with time varying series and shows good extrapolation properties [10,11,12,13]. But this approach needs, for the training, large amount of data, i.e. 3D kinematics human data and, to achieve a good result, a very high number of units.

Nowadays computer vision techniques provide tools to reconstruct three-dimensional trajectories of selected body landmarks, suitably marked [14]. Due to the high complexity of the human kinematics, only simplified models can be adopted, which represent the body as a set of linked anatomical segments modeled as rigid bodies. Under this limitation, the required kinematics human data for NN training can be collected through data processing of the 3D coordinates of the captured human motion. Such an approach has been followed here.

2 Methods

The method is based on the definition of a bio-mechanical model matched to a human subject performing motor tasks. A predefined set of marker points on the subject have been tracked and from their trajectories a set of kinematic parameters of the model have been computed to obtain the variables for the neural networks training (3D trajectories and joint angles).

2.1 Biomechanical model

Reaching and pointing targets in sitting posture are typical tasks which involve the movement of the upper part of the body, while the kinematics of the legs, can be neglected. Accordingly, our bio-mechanical model comprises the pelvis and considers the following upper body districts: trunk, head, shoulders, arms and hands for a total of 11 anatomical districts, 11 joints and 30 Dof (see Fig. 1). Each district has been considered as a rigid body and characterized by the anatomical length and a local reference system. The assumed model is required to take into account for the deterministic part of the process that can not be adequately modeled by the conceived NN. The pelvis represents the root of the kinematic chain and is characterized by 6 Dof. Then the other districts are characterized by 3 rotational Dof, except for the clavicles (2 Dof) and the forearms (1 single Dof: the elbow joint permits only the flexion).

In order to compute the anatomical district kinematics, 26 retro-reflective markers have been attached to the subject and acquired by a dedicated hardware [15] during motor task (see Fig.2).

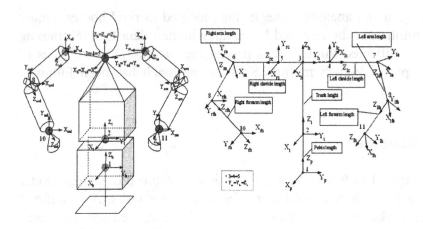

Fig. 1. The biomechanical model.

2.2 Experimental protocol

The subject, seating in front of a desk, had to perform, with the left hand, point to point reaching movements starting from points A and B and ending to points C, D and E, all lying in the table (see Fig. 2) and move back to the corresponding starting position. The 2nd metacarpal head was considered as end-effector. Neither postural nor kinematics as well as dynamics constraints were imposed to the subject. Each movement was repeated 40 times, for a total of 240 files, and the duration of each repetition was about 2 seconds.

2.3 Data processing

The bio-mechanical model has been designed to be controlled by the rotation angles expressed in local reference systems. The hypothesis on the model is that accurate structural features can be estimated from the acquired data. Due to the skin motion and the measurement errors, the anthropometrical measures can vary frame by frame. A least square estimate of those values have been computed as the mean of the distribution of the corresponding body segment lengths computed during the motion. Nevertheless this assumption does not allow well fitting the real trajectories of the acquired markers.

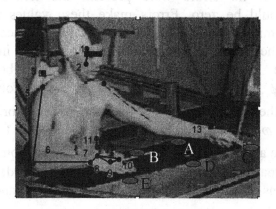

Fig. 2. The subject performing the task.

Actually, by using the mean anthropometrical lengths and keeping unchanged the rotational part of each transformation matrix, positioning errors are observed at each district and they propagate through the chain making more and more relevant their effect when approaching the hands.

Since our goal was to achieve high accuracy in controlling the end-effectors, an ad-hoc correction procedure has been implemented which reduces the error between the position of the end-effector of the model and of the human subject based on the following cost function J:

$$J = \min\left(w_1\left(P_{2h} - MP_{2b}\right)^2 + w_2\left(P_{4h} - MP_{4b}\right)^2 + w_3\left(P_{wh} - MP_{wb}\right)^2 + w_4\left(V_h - RV_b\right)^2 + c\right) \quad \textbf{(1a)}$$

$$c = w_5\left(\partial\omega_c^2 + \partial\phi_c^2\right) + w_6\left(\partial\omega_h^2 + \partial\phi_h^2 + \partial\kappa_h^2 + \partial\phi_{fa}^2 + \partial\omega_a^2 + \partial\phi_a^2 + \partial\kappa_a^2\right) \quad \textbf{(1b)}$$

where P_{2h}, P_{4h}, P_{wh} and V_h are respectively the 3D coordinates of the 2nd and the 4th metacarpal head, of the wrist rotation center and the direction cosine of the normal to the plane π expressed in the trunk reference frame, while P_{2b}, P_{4b}, P_{wb} and V_b are the 3D coordinates of the same points and the direction cosine of the normal expressed in the hand local reference frame; M is the roto-translation matrix and R is the rotation matrix between the trunk and the hand reference system. Thus, if no errors were present, the terms between parentheses should be zero. Errors make these values non zero requiring their minimization. The c term (see Eq. 1b) accounts for the control of the angle variations during the correction in order to constrain the norm of the solution to be small ($\partial\omega_c, \partial\phi_c, \partial\omega_h, \partial\phi_h, \partial\kappa_h, \partial\phi_{fa}, \partial\omega_a, \partial\phi_a, \partial\kappa_a$ are the correction to be performed on the clavicle (c), hand (h), arm (a) and forearm (fa) rotation angles ω-yaw. ϕ-pitch, κ-roll). The weights w_1 to w_4 ($w_1 = w_2 = w_3 \geq w_4$) are adjusted in order to get the best repositioning of the second metacarpal head, while w_5 and w_6 ($w_6 \geq w_5$) in order to have major corrections on the angles of the arm, forearm and hand.

2.4 Neural network architecture

We focused our simulation only on the arm complex (arm + forearm + hand in the bio-mechanical model) characterized by 7 Dof (three rotation angles of the shoulder, one rotation angle of the elbow and the three rotation angles of the wrist). Let notice that the shoulder is not fixed at ground so we should take into account for its 3D coordinates. This raises the number of the involved variables to 10, but, as will be seen in the following, the knowledge of the shoulder coordinates in time, allows using 7 Dof only. Our effort was first oriented towards the choice of the proper input/output data set to be used by the NN architecture in the training stage, through a selection process based on the stability of the resulting trajectories of the wrist. This choice was dictated by the variability and the excursion of the wrist which is greater than all the other points of arm complex, but more predictable than the metacarpal heads. Since NN training based on angular coordinates produced accurate results for angles but very poor for 3D coordinates, a mixed approach (3D coordinates + angle information) has been pursued.

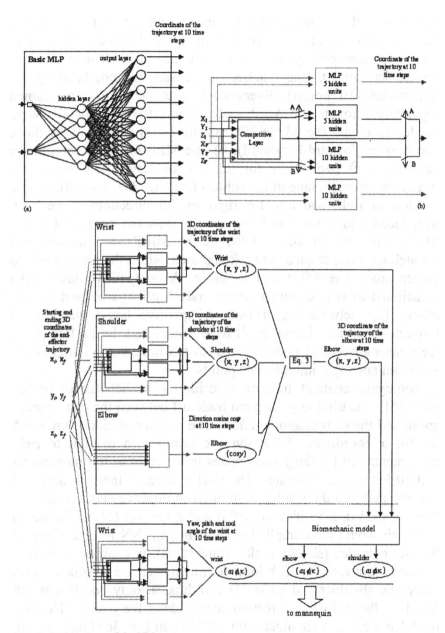

Fig. 3. Architecture of the NN system

The kernel of the network is a multilayer perceptron (MLP) (see Fig. 3a). Its input is one of the 3D coordinates (*x*, or *y*, or *z*) of the point from which the subject started the movements (A or B in figure 2) and the target point (C or D or E in figure 2), which the subject reached before getting back to the starting point. The number of outputs is equal to the number of the sampled points along the trajectories (10 points). Every coordinate (x, y, z) was trained separately through one MLP each. Four different networks were needed to obtain the whole 3D trajectories: a) one MLP for the x coordinates (forward-backward); b) one MLP for the z coordinates (up-down); c) two MLPs for the y coordinates (left-right) with respect the frontal plane of the subject (see Fig. 3b). Two MLPs, one for left to right and one for right to left directions, have been introduced which provide better results than the case of a single MLP. Thus an automatic left-right classifier was implemented through an unsupervised network and included in the architecture to switch from one MLP and to other. A competitive layer (CL) constituted by two Kohonen neurons (see Fig. 3b) was used for this block. The network (apart from competitive layer) was trained through supervised learning; all trajectories used during this phase were normalized and undersampled to ten time steps: initial point, end point and eight intermediate points.

A conjugate gradient training algorithm (Fletcher-Reeves update rule [16]) was used to get a good trade off between the convergence speed and the computational effort. The number of hidden units is 5 for the x coordinate, 5 for the y coordinate in the left to right movements, 10 for the y coordinates in the right to left movements and 10 for the z coordinates. The total number of motor tasks (240) was divided into three subsets, in order to improve generalization: a training (85%), a validation (10%) and a test set (5%). The trivial approach would have implied to implement a NN for each joint of the arm complex (shoulder, elbow and wrist). In contrast, instead of training a network for each joint coordinate set, we considered that, given the shoulder and wrist 3D coordinates, only one Dof is left (i.e. the elbow flexion, direction cosine along the y axis). Thus we modeled wrist and shoulder with a NN as in Fig. 3c (final network for the whole arm movement generation) and implemented a pre-processing algorithm aimed to determine the elbow Dof. A two-layer perceptron has been used to model the variable. The 3D coordinates of the shoulder, elbow and wrist have been used to compute the angles of the shoulder and the elbow according to the bio-mechanical model. In order to model also the end-effector (2nd metacarpal head), we trained another network on the three wrist rotation angles. At the end ten variables were necessary to the animation of the arm complex: the 3D coordinates of the shoulder, the direction cosine of r line, the 3D coordinates of the wrist and the three rotation angles of the wrist. Finally, to assess the reliability of the variables used for the NN training, we tried the trivial approach by training a network directly on the 7 joint angles of the arm.

3 Results

Table 1 reports the RMS error of the elbow coordinates obtained by training its coordinates (3D trajectory). It should be noticed that the choice of the y direction cosine guarantees the reliability of anatomical segments lengths while the errors on 3D trajectory of the elbow reflects on a variation of these, making the whole procedure unsuitable for the use with CAD mannequins.

Elbow kinematics	x (cm)	y (cm)	z (cm)
3D coordinates	1.4 ± 0.66	2.5 ± 2.01	1.1 ± 0.62
Direction cosine	1.5 ± 0.68	1.6 ± 1.06	1.09 ± 0.4

Table 1. The RMS error of the elbow coordinates obtained by training its coordinates.

The reliability of the whole architecture has been evaluated on the trajectory of the end-effector. In table 2, the root mean square error between the real and network estimated trajectories of the end-effector is reported, together with the variability of the real trajectory (standard deviation on the real trajectory used by the subject). These results were compared with those obtained by training the network directly on the joint angles. As expected the results turned out to be definitely worse because of error propagation along the kinematic chain.

RMSE	x (cm)		y (cm)		z (cm)	
	TRAINING SET	TEST SET	TRAINING SET	TEST SET	TRAINING SET	TEST SET
END EFFECTOR	2.06±1.01	2.53±0.7	0.91±1.42	3.01±1.18	1.62±0.63	2.01±0.61
VARIABILITY	1.52		0.81		0.75	

Table 2. The RMSE between the real and estimated trajectories of the end-effector

4 Discussion and Conclusions

The aim of this work was to develop a framework to reproduce accurate human motion trajectories with the ultimate goal of controlling a robot manipulator moving in a human like way. The choice to adopt a NN-based approach instead of a less blind model, much is known of the underlying biomechanical system and it would not seem hard to design and manage it, was motivated by the two following problems: accuracy int the estimation of the structural parameters and redundancy of the degrees of freedom of the involved articulated structure (inverse kinematics). Actually, due to anatomical complexity, structural parameters can be estimated only with a finite accuracy, that can distort the prediction capability of the kinematic models. Moreover, approaches using direct and inverse kinematics, are not able alone to reproduce accurate trajectory (high resemblance with real human trajectories) and additional rules (dynamics/energy) have to be added to contraints redundant. The selected approach has required the implementation of a pre-processing procedure to accurately estimate the lengths of the anatomical parts, - the implemented optimization comes to play a central role because it establishes some clear-cut hypotheses about the degrees of freedom of the mannequin and the nature of the raw data before any kind of processing takes place. Furthermore, our choice considers the compatibility with the different models used till now by the different CAD system [1,2,3,4]. To the second purpose NN approach allows removing the need of additional contraint but a high number of training trajectories is mandatory. NN is not new to solve this problem. In 1998, Steinkühler et al. [17] proposed a MMC (Mean value of Multiple Computation) net for the control of a multi-segmented manipulator with redundant degrees of freedom. The problem of this approach is that the computed end-effector paths are different from those showed by arm movements: they do neither show straight paths nor bell-shaped profile. In contrast our approach, being based on a real human data, does not present these problems and allows animating the arm complex specifying the 3D coordinates of the starting and ending position of the end-effector; the neural network outputs constitutes the input to the bio-mechanical model animation. The results (table 1-2) show a fairly good accuracy in the simulation on the human movements, especially if we consider human variability [18]. Future work will be oriented to the animation of the whole mannequin, using a suitable number of neural networks for its animation. The splitting of the kinematic chain in some key points that will prove to reduce error propagation problems will be extended to the remaining biomechanical model. Thanks to the good generalization properties, this approach can be useful to generate human like trajectories of any kind of point to point movement. More complex movements, like obstacle avoidance, could be generated by a neural network by suitably splitting the trajectories and obviously, by using a more complex architecture.

References

[1]. N.I. Badler, "Human factors simulation research at the University of Pennsylvania. Computer graphics research quarterly progress report No. 38," Department of Computer and Information Science. University of Pennsylvania. Fourth Quarter 1-17, 1990.

[2]. P. Bapu, M. Korna and J. McDaniel, "User's guide for COMBIMAN programs (COMputerized Biomechanical MAN-model)," AFAMRL-Tech. Report pp. 83-97, 1983.

[3]. E.J. Szmrecsanyl, "CYBERMAN – An interactive computer graphics manikin for human factors analysis," Control Data Corporation. Worldwide User Conference.

[4]. M. Cavazza, R. Earnshaw, N. Magnenat-Thalmann and D. Thalmann, "Motion control of virtual humans," *IEEE Computer Graphics and Applications*, vol. 18, pp. 24-31, 1998.

58

[5]. H. Cruse, U. Steinkühler, "Solution of the direct and inverse kinamatic problems by a common algorithm based on the mean of multiple computations," *Biol. Cybern.*, vol. 69, pp. 345-351, 1993.

[6]. H. Cruse, "Constraints for joint angle control of the human arm," *Biol. Cybern.*, vol. 54, pp. 125-132, 1986.

[7]. T. Yoshikawa, "Manipulatability and redundancy of robotic mechanism," *Proceeding of the 1985 IEEE,* Computer Society Press, Silver Spring, Md, 1004-1009.

[8]. T. Flash, N. Hogan, "The co-ordination of the arm movements: an experimentally confirmed mathematical model," *Journal of neuroscience,* vol. 7, pp. 1688-1703, 1985.

[9]. Y. Uno, M. Kawato, R. Suzuki, "Formation and control of optimal trajectory in human arm movement," *Biol. Cybern.*, 61, pp. 89-101, 1989.

[10]. M. Kawato, "Computational schemes and neural network models for formation and control of multijoint arm trajectory," In: *Neural Network for control* ed. W.T. Miller III, R. Sutton and P. Werbos (1990), MIT Press.

[11]. M. Kawato, M. Isobe, Y. Uno and R. Suzuki, "Trajectory formation of arm movements by cascade neural network model based on minimum torque-change criterion," *Biol Cybern.,* vol. 62, pp. 275-288, 1990.

[12]. J. C. Fiala, "A network for learning kinematics with application to human reaching models," *Proc. of the ICNN*, 5, pp. 2759-2764, Orlando, FL, 1994.

[13]. P. Morasso and V. Sanguineti, "Self-organizing body-schema for motor planning," *J. Motor Behavior*, vol. 26, pp. 131-148, 1995.

[14]. A. Pedotti and G. Ferrigno, "Opto-electronic based system," in *Three dimensional analysis of human movement*, P. Allard, I.A.F. Stokes and J.P. Bianchi, pp. 57-77, Human Kinetics Publishers, 1995.

[15]. F. Lacquaniti, G. Ferrigno, A. Pedotti, J.F. Soechting and C. Terzuolo, "Changes in spatial scale in drawing and handwriting contributions by proximal and distal joints," *The journal of Neuroscience,* vol. 7, no. 3, 1987.

[16]. Fletcher, R., and C.M. Reeves, "Function minimization by conjugate gradients," *Compute Journal*, vol. 7, pp. 149-154, 1964.

[17]. U. Steinkühler, H Cruse, "A holistic model for an internal representation to control the movement of a manipulator with redundant degrees of freedom," *Biol. Cybern.*, vol. 79, pp. 457-466, 1998.

[18]. N.A. Bernstein, "The co-ordination and regulation of movements," *Pergamon Press*, New York, 1967.

A New ANFIS Synthesis Approach for Time Series Forecasting

Massimo Panella, Fabio Massimo Frattale Mascioli,
Antonello Rizzi, and Giuseppe Martinelli

INFO-COM Department, University of Rome "La Sapienza",
Via Eudossiana 18, I-00184 Rome, Italy
E-mail: panella@infocom.uniroma1.it
Home page: http://infocom.uniroma1.it/~panella

Abstract: ANFIS networks are neural models particularly suited to the solution of time series forecasting problems, which can be considered as function approximation problems whose inputs are determined by using past samples of the sequence to be predicted. In this context, clustering procedures represent a straightforward approach to the synthesis of ANFIS networks. The use of a clustering procedure, working in the conjunct input-output space of data, is proposed in the paper. Simulation tests and comparisons with other prediction techniques are discussed for validating the proposed synthesis approach. In particular, we consider the prediction of environmental data sequences, which are often characterized by a chaotic behavior. Consequently, well-known embedding techniques are used for solving the forecasting problems by means of ANFIS networks.

1 Introduction

The prediction of future values of real-world data sequences is often mandatory to the cost-effective management of available resources. Consequently, many worldwide research activities are intended to improve the accuracy of well-known prediction models. Among them, an important role can be played by Adaptive Neuro-Fuzzy Inference Systems (ANFIS) [1]. In fact, as it will be pointed out in the paper, such computational models can be used as predictors by means of a suitable transformation of the prediction problem into a function approximation one.

ANFIS network performs the approximation of an unknown mapping $y=f(\underline{x})$, $f:\Re^N \rightarrow \Re$, by implementing a fuzzy inference system constituted by M rules of Sugeno first-order type. The k-th rule, k=1...M, has the form:

$$if\ x_1\ is\ B_1^{(k)},\ and\ x_2\ is\ B_2^{(k)},...,\ and\ x_N\ is\ B_N^{(k)}\ then\ y^{(k)} = \sum_{j=1}^{N} a_j^{(k)} x_j + a_0^{(k)} \quad (1)$$

where $\underline{x} = [x_1\ x_2\ ...\ x_N]^t$ is the input pattern (with 't' denoting transposition), and $y^{(k)}$ is the output associated with the rule. The rule is characterized by the membership functions (MFs) of the fuzzy input variables $B_j^{(k)}$, j=1...N, and by the coefficients $a_j^{(k)}$, j=0...N, of the crisp output. Several alternatives are possible for choosing the fuzzification type of crisp inputs, the composition of input MFs, and the way rule outputs are combined [1].

When dealing with data driven estimation of the mapping $y=f(\underline{x})$, the latter is known by means of numerical examples, i.e. by a training set of P input-output pairs $\{\underline{x}_i, y_i\}$, i=1...P. In this case, a useful approach to the synthesis of ANFIS networks is based on clustering the training set. Different types of clustering approaches can be used in this regard [2]. For instance, clustering in the conjunct input-output space can overcome some drawbacks pertaining most of traditional approaches, where clustering is used only for determining the rule antecedents in the input space [3].

We present in Sect. 2 an ANFIS synthesis procedure that is based on the conjunct input-output space approach. Firstly, we determine the coefficients of the Sugeno rule consequents by using a suitable clustering procedure. Then, we use a fuzzy classifier in order to determine the fuzzy variables of Sugeno rule antecedents. Several fuzzy Min-Max classifiers can be used for this purpose [4]; a well-known class is based on adaptive resolution mechanisms [5], which is overviewed in Sect. 3.

The generalization capability of ANFIS networks can be satisfactory, provided that their architectures consist of a suitable number of rules. This is a crucial problem since these neuro-fuzzy networks can be easily overfitted in case of noisy or ill-conditioned data. In order to determine automatically the optimal number of rules, we also suggest in Sect. 2 an optimization of the proposed synthesis procedure based on basic concepts of learning theory.

As previously mentioned, the use of ANFIS networks may result very effective also in several prediction problems. In fact, the latter can be considered as function approximation problems whose inputs are suitably determined by using past samples of the sequence to be predicted. We illustrate in Sect. 4 a standard technique, which is particularly suited to the prediction of real-world data sequences that often manifest chaotic behaviors.

The validity of the proposed ANFIS synthesis procedure is ascertained in Sect. 5, where several benchmark results, obtained by using environmental data sequences, are presented.

2 The Hyperplane Clustering Synthesis Procedure

The use of clustering in ANFIS synthesis reduces the redundancy in the data space(s), allowing the determination of significant rules directly from the clusters that have been discovered in the available data set. Each rule corresponds to a structured set of points in these spaces. There are possible 3 types of data spaces for clustering: input, output, and conjunct input-output [2].

Traditional techniques, either based on input or output space, critically depend on the regularity properties of the mapping to be approximated. In fact, such techniques implicitly assume that close points in the input space (belonging to the same rule) are mapped into close points in the output space. Unfortunately, this not an usual situation, especially in forecasting real-world data sequences where the mapping to be approximated might present several irregularities and sharp behaviors.

In order to overcome the problems of the previous approaches, we illustrate in the following the use of a clustering strategy based on a conjunct input-output space [3]. Such a clustering procedure is oriented to the direct determination of the structure of the unknown mapping $f(\underline{x})$. Namely, the ANFIS architecture can be considered as a piecewise linear regression model, where $f(\underline{x})$ is approximated by a suitable set of M hyperplanes, each related to an input-output cluster. Therefore, the prototype of the k-th cluster, k=1...M, in the conjunct input-output space will be represented by the coefficients $a_j^{(k)}$, j=0...N, which determine the linear consequent of the corresponding k-th rule. We propose an alternating optimization technique, i.e. a C-means clustering in the 'hyperplane space', in order to determine such prototypes:

- *Initialization*. Given a value of M, the coefficients of each hyperplane are initialized by following a suitable criterion. In this case, we choose a simple random initialization of them. Successively, each pair $\{\underline{x}_i, y_i\}$, i=1...P, of the training set is assigned to the hyperplane A_q, $1\leq q\leq M$, based on the procedure discussed in Step 2.

- *Step 1*. The coefficients of each hyperplane are updated by using the pairs assigned to it (either in the successive step 2 or in initialization). For the k-th hyperplane, k=1...M, a set of linear equations has to be solved:

$$y_t = \sum_{j=1}^{N} a_j^{(k)} x_{tj} + a_0^{(k)} \tag{2}$$

where index 't' spans all pairs assigned to the k-th hyperplane. Any least-square technique can be used to solve the previous set of linear equations.

- *Step 2*. Each pair $\{x_i, y_i\}$ of the training set is assigned to the updated hyperplane A_q, with q such that:

$$e_i = \left| y_i - \left(\sum_{j=1}^{N} a_j^{(q)} x_{ij} + a_0^{(q)} \right) \right| = \min_{k=1...M} \left| y_i - \left(\sum_{j=1}^{N} a_j^{(k)} x_{ij} + a_0^{(k)} \right) \right| \tag{3}$$

- *Stop criterion*. If the overall error defined by

$$E = \sum_{i=1}^{P} e_i \tag{4}$$

has converged then stop, else go to step 1.

The previous algorithm only yields the linear consequent of Sugeno rules. However, each hyperplane may correspond to several clusters in the input space, i.e. to well separated sets of input patterns $\underline{x}_t$ associated with it. Consequently, the determination of the input MFs is not straightforward. This problem can be solved by considering a suitable classification problem, where each pattern is labeled with an integer q, $1 \leq q \leq M$, representing the hyperplane to which it has been previously assigned. Any fuzzy classification algorithm can be used in this regard. The use of the Adaptive Resolution Classifier (ARC), belonging to the class of well-known Simpons's Min-Max models, is suggested in the next section.

The combination of both the hyperplane clustering and the classification in the input space, for a given value of M, allows to determine the ANFIS network. This procedure will be denoted in the following as Hyperplane Clustering Synthesis (HCS) algorithm. The HCS algorithm, and the resulting ANFIS network, depends upon a given value of M. The optimal value of M, yielding the best performing network in terms of generalization capability (i.e. lowest error on a test set), should be accomplished during training without any knowledge about the test set. We proposed in [3] a constructive technique, denoted as Optimized HCS (OHCS), which is the most general optimization technique that can be pursued in this context. The value of M is progressively increased and several ANFIS networks are generated, by using the HCS algorithm, in correspondence to any M. In fact, HCS uses a random initialization of the hyperplanes, hence different initializations can yield different networks for the same value of M.

Successively, the optimal value of M is chosen by relying on basic concepts of learning theory [6], i.e. by finding the minimum value of the following cost functional:

$$F(M) = (1 - \lambda) \frac{E(M) - E_{min}}{E_{max} - E_{min}} + \lambda \frac{M}{P} \tag{5}$$

where E(M) is the training set performance (i.e. the approximation accuracy on the training set); E_{min} and E_{max} are the min and max values of E, obtained during the investigation of different values of M; λ is a weight in the range [0, 1]. This weight is not critical since the results are slightly affected by the variation of it in a large interval centered in 0.5.

3 The Use of Adaptive Resolution Classifiers for the Determination of Antecedent MF's

Min-Max classification technique consists in covering the patterns of the training set with hyperboxes (HBs). It is possible to establish size and position of each HB by two extreme points: the 'Min' and 'Max' vertices. The hyperbox can be considered as a crisp frame on which different types of membership functions can be adapted. In the following, we will adopt the original Simpson's membership function [4], in which the slope outside the hyperbox is established by the value of a parameter γ.

The original training algorithm, and almost all its successive modified versions, are characterized by the following drawbacks: an excessive dependence on the presentation order of training set; the same constraint on covering resolution in the overall input space. In order to overcome these inconveniences, a new training algorithm has been developed in [5]: the Adaptive Resolution Classifier (ARC).

The basic operation for obtaining a variable resolution is the 'hyperbox cut'. It consists in cutting a hyperbox in two parts by a hyperplane, perpendicularly to one coordinate axis and in correspondence to a suitable point. The underlying purpose of these cuts is to generate hyperboxes that cover a set of patterns associated with only one class. These hyperboxes will be denoted as 'pure', in contrast with the others denoted as 'hybrid'. In order to comply with the principles of learning theory, pure hyperboxes belonging to the same classes are fused if no overlaps result with both hybrid hyperboxes and pure hyperboxes of different classes. Hyperboxes resulting from cutting operation are contracted to the minimum size necessary for containing all their patterns.

The generic step of ARC training algorithm consists in selecting one hybrid hyperbox and then cutting it. If one (or both) pure hyperbox has been created, the algorithm tries to fuse it with other pure hyperboxes associated with the same class, in order to reduce the overall complexity of the classifier.

During the training process, the algorithm generates a succession of Min-Max classifiers characterized by an increasing complexity (number of neurons of the hidden layer) and classification accuracy on the training set. According to learning theory, the optimal Simpson's classifier is obtained in correspondence to the minimum value of a suitable functional cost, which is similar to that proposed in (5).

As shown in [5], ARC algorithm outperforms the original Min-Max algorithm, and some optimized versions of it, both in the generalization capability and in the training time. Moreover, Simpson's classifiers trained by ARC are independent of pattern presentation order. Therefore, on the basis of these results, we adopted the ARC algorithm in order to find the input MFs of the ANFIS network.

As previously stated, several HBs of the optimal Simpons's classifier can be associated with the same hyperplane (rule) determined by the hyperplane clustering. In this case, the overall input MF $\mu_{\underline{B}(k)}(\underline{x})$, k=1...M, of each rule can be determined on the basis of the composition operators usually adopted for fuzzy Min-Max neural networks [4]. For instance, if $H_1^{(q)}, H_2^{(q)}, \ldots, H_R^{(q)}$ are the HBs associated with the class label q, and $\mu_1^{(q)}(\underline{x}), \mu_2^{(q)}(\underline{x}), \ldots, \mu_R^{(q)}(\underline{x})$ the corresponding MFs, then we will have:

$$\mu_{\underline{B}(q)}(\underline{x}) = \max\left\{\mu_1^{(q)}(\underline{x}), \mu_2^{(q)}(\underline{x}), \ldots, \mu_R^{(q)}(\underline{x})\right\} \qquad (6)$$

4 The Function Approximation Approach for Time Series Forecasting

Due to the actual importance of forecasting, the technical literature is plenty of proposed methods for implementing a predictor, especially in the field of neural networks [7]. The general approach for solving a prediction problem is based on the solution of suitable function approximation problem. Let us consider a general predictor, which is used to predict a sampled sequence S(t), and a general function approximation problem y=f($\underline{x}$), f:$\Re^N \rightarrow \Re$. For example, the simplest prediction approach is based on linear autoregressive (AR) models; each input vector x_t is constituted by N consecutive samples of S(t) and the target output y_t is the sample to be predicted at a distance m:

$$\underline{x}_t = [S(t) \ S(t-1) \ ... \ S(t-N+1)] \ , \quad y_t = S(t+m)$$

$$f_{AR}(\cdot) = -\sum_{j=1}^{N} \lambda_j x_{tj} \ \Rightarrow \ \hat{S}(t+m) = -\sum_{j=1}^{N} \lambda_j S(t-j+1) \qquad (7)$$

where $\hat{S}(t+m)$ denotes the estimation of the actual value $S(t+m)$.

Usually, the way to determine the input vectors $\underline{x}_t$, based on past samples of S(t), is called embedding technique. The function $f_{AR}(\cdot)$, i.e. the coefficients λ_j, j=1...N, can be determined in this case by relying on global statistical properties of the sequence S(t), i.e. on its autocorrelation function.

Real-world data sequences often posses a chaotic behavior that is typical for almost all real-world systems. The accurate prediction of future values of such sequences is often mandatory to the cost-effective management of available resources. The performance of a predictor depends on how accurate it models the unknown context delivering the sequence to be predicted. Unfortunately, when dealing with chaotic sequences, the previous AR predictor would hardly fail since it is based on a linear approximation model and on trivial embedding technique. Consequently, more care should be taken on choosing both the approximation model and the embedding parameters.

In the case of a chaotic sequence S(t), the latter can be considered as the output of a chaotic system that is observable only through S(t). Consequently, the sequence S(t) should be embedded in order to reconstruct the state-space evolution of this system, where $f(\cdot)$ is the function that approximates the relationship between the reconstructed state (i.e. $\underline{x}_t$) and its corresponding output (i.e. the value y_t to be predicted) [8]. Because of the intrinsic non-linearity and non-stationarity of a chaotic system, $f(\cdot)$ should be a non-linear function, which can be determined only by using data driven techniques.

The usual embedding technique, which is useful for chaotic sequences, is based on the determination of both the embedding dimension D of the reconstructed state-space attractor and the time lag T between the embedded past samples of S(t); i.e.:

$$\underline{x}_t = \left[S(t) \ S(t-T) \ S(t-2T) \ ... \ S(t-(D-1)T) \right]^t \tag{8}$$

where the superscript 't' denotes transposition. Both the values of D and T will be determined in the following by applying the methods suggested in [8]. In particular, D will be obtained by using the False Nearest Neighbors (FNN) method, whereas T will be obtained by using the Average Mutual Information (AMI) method.

From the above discussion, it follows that the implementation of a predictor will coincide with the determination of a non-linear data driven function approximation model $y_t = f(\underline{x}_t)$. For this purpose, we propose in this paper the use of ANFIS networks. In fact, we will demonstrate in the next section that the robustness of fuzzy logic in approximating also very irregular regions of the mapping, along with the optimization of the structure complexity, provide to the ANFIS networks both efficacy and flexibility in solving such forecasting problems.

5 Illustrative Tests

The forecasting performances of ANFIS networks, resulting from the proposed OHCS procedure, have been carefully investigated by the several simulation tests we carried out in this regard. We will illustrate in the following the results concerning actual environmental data sequences. They consist on the observation of some pollution agents and of the electric power consumption in the downtown of Rome (Italy).

The OHCS procedure is evaluated by comparing the prediction performance of the resulting ANFIS network with respect to the performances obtained by using two different predictors: the linear AR predictor introduced in (7) and determined by the well-known Levinson-Durbin algorithm; another ANFIS predictor whose network is generated in this case by applying the Subtractive Clustering (SUBCL) method for rule extraction [9] and then the standard least-squares method together with the back-propagation gradient [1].

All the previous computational models, which will implement the function approximator $y_t = f(\underline{x}_t)$, are trained on the first 2000 samples of S(t). These samples are also applied to both the AMI and FNN methods, in order to compute the embedding dimension and the time lag. The performances of the resulting predictors are tested on the successive 600 samples of the sequence (slight changes may occur in these numbers because of the different embedding quantities used for each sequence). The performance is measured by the Normalized Mean Squared Error (NMSE) defined as the ratio between the mean squared prediction error and the variance of the sequence to be predicted.

The first two sequences of the following tests are obtained by measuring the level of Ozone (in μg) and of acoustic noise (in dB) in a given location of the downtown of Rome (Italy). Such sequences are obtained by using a sampling rate of 5 minutes. The third sequence is the level of the electric power consumption in (MW) measured in Rome. Its values are obtained by using a sampling rate of 1 hour. The results are summarized in Tab. 1: the first three columns indicate, respectively, the time lag (T), the embedding dimension (D), and the number of ANFIS rules (M) determined by the OHCS procedure; the successive three columns show the NMSEs obtained by using the predictors under investigation.

Test	T	D	M	OHCS	AR	SUBCL
Ozone	3	5	16	$1.72 \cdot 10^{-1}$	$2.43 \cdot 10^{-1}$	$2.19 \cdot 10^{-1}$
Acoustic noise	4	14	9	$3.47 \cdot 10^{-1}$	$4.39 \cdot 10^{-1}$	$3.04 \cdot 10^{-1}$
Electric load	7	5	13	$9.93 \cdot 10^{-3}$	$4.95 \cdot 10^{-2}$	$3.45 \cdot 10^{-2}$

Table 1. Prediction results on real data sequences

6 Conclusions

ANFIS networks are neural models particularly suited to the solution of function approximation problems. We proposed in this paper the use of ANFIS networks for solving forecasting problems based on well-known techniques for chaotic system modeling.

The ANFIS synthesis is carried out in the paper by means of the HCS procedure, which is based on the conjunct input-output space data clustering. HCS algorithm determines the consequent part of each first-order Sugeno rule by using a C-means clustering in the hyperplane space, then it finds the corresponding antecedent part by means of the ARC fuzzy Min-Max classifier. The ANFIS network is determined by using an optimization procedure (i.e. OHCS) of the HCS algorithm, so that the optimal number of rules, and therefore the best generalization capability of the network, is automatically achieved.

ANFIS networks obtained by using the OHCS procedure are used in this paper for forecasting real-world data sequences. The results are encouraging in extending the proposed synthesis procedure in critical forecasting applications, as for industrial and environmental ones, where even slight improvements of the prediction accuracy might result in a more effective management of the human and economic available resources.

References

1. Jang, J.S., Sun, C.T., Mizutani, E.: Neuro-Fuzzy and Soft Computing: a Computational Approach to Learning and Machine Intelligence. Prentice Hall, Upper Saddle River, NJ USA (1997)

2. Frattale Mascioli, F.M., Mancini, A., Rizzi, A., Panella, M., Martinelli, G.: Neurofuzzy Approximator based on Mamdani's Model. Proc. of WIRN2001, Vietri Sul Mare, Salerno, Italy (2001)

3. Panella, M., Rizzi, A., Frattale Mascioli, F.M., Martinelli, G.: ANFIS Synthesis by Hyperplane Clustering. Proc. of IFSA/NAFIPS 2001, Vancouver, Canada (2001)

4. Simpson, P.K.: Fuzzy Min-Max Neural Networks–Part 1: Classification. IEEE Transactions on Neural Networks, Vol. 3, No. 5 (1992) 776-786

5. Rizzi, A., Panella, M., Frattale Mascioli, F.M.: Adaptive Resolution Min-Max Classifiers. To appear in IEEE Transactions on Neural Networks (2001-2002)

6. Haykin, S.: Neural Networks, A Comprehensive Foundation. Macmillan, New York, NY (1994)

7. Masulli, F., Studer, L.: Time Series Forecasting and Neural Networks. Invited tutorial in Proc. of IJCNN'99, Washington D.C., USA (1999)

8. Abarbanel, H.D.I.: Analysis of Observed Chaotic Data. Springer-Verlag, Berlin Heidelberg New York (1996)

9. Chiu, S.: Fuzzy Model Identification Based on Cluster Estimation. Journal of Intelligent & Fuzzy Systems, Vol. 2, No. 3 (1994)

Experiments on a Prey Predators System

V.Di Gesù, G.Lo Bosco

Dipartimento di Matematica ed Applicazioni, Univ.Palermo – Via Archirafi 34, 90123 Palermo, Italy (digesu,lobosco@math.unipa.it)

Astract: The paper describes a prey-predators system devoted to perform experiments on concurrent complex environment. The problem has be treated as an optimization problem. The prey goal is to escape from the predators reaching its lair, while predators want to capture the prey. At the end of the 19th century, Pareto found an optimal solutions for decision problems regarding more than one criterion at the same time. In most cases this 'Pareto-set' cannot be determined analytically or the computation time could be exponential. In such cases, evolutionary Algorithms (EA) are powerful optimization tools capable of finding optimal solutions of multi-modal problems. Here, both prey and predators learn into an unknown environment by means of genetic algorithms (GA) with memory. A set of trajectories, generated by a GA, are able to build a description of the external scene driving a predators to a prey and the prey to the lair. The prey-predator optimal strategies are based on field of forces. This approach could be applied to the autonomous robot navigation in risky or inaccessible environments (monitoring of atomic power plants, exploration of sea bottom, and space missions).

Keywords:genetic algorithms, optimization, autonomous navigation, prey-predators systems.

1 Introduction

Recently, studies, refining or extending the Lotka Volterra model [1,2], have been developed on the influence of predators on preys to study the instability of the natural ecosystems and to models the ecosystem around [3].

In this paper a prey-predators system is described; it has been designed in order to perform experiments on complex environment. The prey goal is to escape from the predators reaching its lair, while the predators goal is to capture the prey. Therefore, the problem can be described as an optimization problem where there are conflicting objectives. At the end of the 19th century, Pareto found on optimal solutions for decision problems regarding more than one criterion at the same time. However, Pareto's solutions can not be always determined analytically or, if they exist, the computation time could be exponential.

In the case of multi-modal problems Evolutionary Algorithms (EA's) can be used to find near-optimal solutions. EA's evolve by imitating features of organic evolution [4]. A simulation, based on genetic algorithms (GA's), has been taken for studying the modeling of external environment into the prey and predator context.

In [5], a predator, represented by unit mobile robots, builds an environment description by using both the generated trajectories and a connectionist paradigm for the detour problem solutions. Here, a prey-predators approach is studied in the framework of GA. Two basic features of predator-prey interactions have been considered: i) relationship between the prey's loss and the predator's gain; ii) in absence of the prey, the predator population should die out exponentially [8]. Moreover, predators and prey already know each others and the prey has an internal model of the refuge site (they do not need learning for that). Autonomous robot navigation has been studied in [6] as an application of evolutionary control of the distributed system [7]. Abdul-Aziz Yakubu demonstrates the stable coexistence of species that would otherwise exclude each other without a prey refuge [9].

In our system design we have introduced a relaxed memory mechanism that is used to emulate mental image creation. Fig. 1 shows an example of prey-predators configuration as it appear in the simulation.

The paper will report some experiments regarding the relationship between the memory mechanism and the capability of predators in finding a solution for capturing the prey. In Section 2 the architecture of the prey-predators system is described. In section 3 details are given of the used GA with memory. Section 4 shows experiments on some complex configurations that our model is able to solve. Conclusion will be given in Section 5.

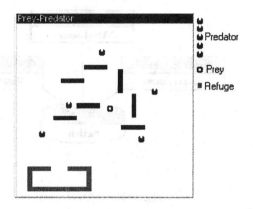

Fig. 1. Prey predator distribution

2 System overview

The system emulates an environment with *n* predators that want to capture one prey (the goal). The prey goal is to escape from the predators and to reach its lair. Prey, predators and lair are constrained in a room with obstacles and they do not know their mutual positions. Predators and prey have a visual system that allows them to see each others; range detectors are used to perceive obstacles. Moreover, they have a mental image system, implemented by a relaxed memory, that allows them to remember for a while the directions of their goals (the prey for the predators and the lair for the prey). Prey and predators act by using three agents that are dedicated to the evaluation of the visual information (*prey recognizer*, the *predator recognizer*, the *lair recognizer*, and the *obstacle evaluator*) a fifth agent (the planner) coordinates the movements (action) of prey and predators, according to an optimization strategy (see Fig. 2). The planner/action module can be further detailed as an evolution phase based on the Active Information Fusion (AIF) loop [8]. The AIF loop consists of five modules: *Observe, Evaluation, Optimization, Choose-Next,* and *Action*.

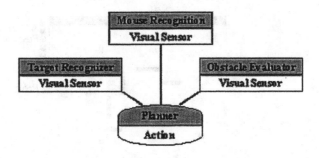

Fig. 2. The ARN cooperating agent

Observe is responsible for the acquisition and preprocessing of *sensory* data. Evaluation compute, on the basis of the observed parameters and the current state of the robot the next choose; the evaluation uses an optimization method (in our case a GA with memory). Choose-Next selects the new movement of the robot on the basis of the evaluation module. It should be noted that all processes may run in parallel mode and they may be executed on the network of distributed processors. The result of the choose module determines the Action (next movement in the environment). After the movement the Observe module drives further sensor-explorations. The Real World represents the environment on which the ARN operates. The information, within the system, flows in a continuous active fusion loop.

3 The proposed GA solution

The planning agent drives predators toward the prey and the prey toward the lair. The problem can be stated, informally, as follows: for each predator i find the minimum number of steps to reach the pray in position $(x_p (t)$, $y_p(t))$ starting from the position $(x_0$, $y_0)$, avoiding obstacles and other predators in the environment. Here t is a time variable that can be considered discrete in the simulation. The problem, before sketched, can be formulated, in the case of one rectangular obstacle of size axb, and one prey in position $(x_p(t)$, $y_p(t))$ as a dynamic programming problem (see Fig. 3):

$$x_i = S_x(\theta_{i-1}, x_{i-1}, y_{i-1}, x_{ob}, y_{ob}, x_F, y_\Gamma; a, b)$$
$$y_i = S_y(\theta_{i-1}, x_{i-1}, y_{i-1}, x_{ob}, y_{ob}, x_F, y_\Gamma; a, b)$$
$$\theta_i = S_\theta(\theta_{i-1}, x_{i-1}, y_{i-1}, x_{ob}, y_{ob}, x_F, y_\Gamma; a, b)$$
$$\min(\delta(x_0, y_0, x_\Gamma, y_\Gamma))$$

The variables involved represent: the coordinates of the predator $(x_i(t), y_i(t))$, the coordinates of the top-left corner of the obstacle (x_{ob}, y_{ob}), the coordinates of the target (x_T, y_T), the view angle θ between the direction of the robot displacement and the vector target. Functions (S_x, S_y, S_θ), describe the dynamical evolution of the system (for sake of simplicity it includes the bounds conditions of the obstacle), the function δ measures the length in step units of the path made by the robot to reach the target. The problem has an exponential complexity, even in the case of a single rectangular obstacle. In real situations, the prey does not know the coordinate system and only the visual sensors and the distance evaluators drive its movements. In our case it can measure the distance from the obstacles and the direction of the target with respect its displacement direction.

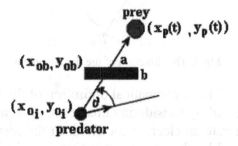

Fig.3. Prey predators assessment

The GA, here described, has been designed to approximate the solution of the, above stated, general optimization problem [9]. The planner, on the basis of the information given by the sensors agents, provides the direction of the displacement. The GA can be sketched as follows:

1. select a starting population of genomes (potential problem solutions) represented as a string of bits;
2. perform the crossover operation on randomly chosen genomes;

3. perform a bit-mutation with probability P to change one or more bit randomly;
4. compute the fitting function, $F(h)$, for each chromosome h;
5. select the new population (with a given technique);
6. if the global fitting, $\overline{F}$, satisfies a given convergence criteria goto 7) else goto 2);
7. end.

Data Representation. One of the key points in GA is the representation of the *genome* on which to perform genetic transformations (selection, crossover and mutation). Our genomes are the displacement directions, θ, in the interval $[-\pi, \pi]$ (see Fig. 4). They are mapped in 16-bits word using the function: $R(\theta) = (\theta + \pi) * (2^{15} - 1) / 2\pi$.

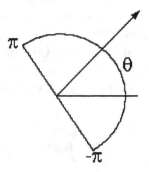

Fig.4. The direction of the displacement.

Fitness Function. The mathematical definition of the fitness function has been formulated by considering obstacles, prey, and predators as entities that generate an electrostatic field. In the case of a predator, the field generated by the prey is attractive (positive) and repulsive (negative) in the case of obstacles and other predators (see Fig. 5.):

$$\overline{F}^{predator\ i} = a_1 \times F_{prey\ i} + a_2 \times \sum_j F_{obstacles\ ij} + a_3 \times \sum_k F_{clash\ ik}$$

where $a_1 + a_2 + a_3 = 1$. While, the prey is repulsed by the predators and attracted by the lair:

$$\overline{F}^{\,prey} = b_1 \times F_{predator_i} + b_2 \times \sum_j F_{obstacles_j} + b_3 \times F_{lair} \; , \; b_1 + b_2 + b_3 = 1$$

Let θ_{ip} be the angle between *predator_i* direction and vector *predator_i* - *prey*, the θ_{pi} angle between *prey* direction and vector *prey-predator_i* , θ_{pl} the angle between *prey* direction and vector *prey - lair*. We define

$$F_{prey_i} = K \times \left(\pi - |\theta_{ip}|\right), \; F_{predator_i} = -H \times \left(\pi - |\theta_{pi}|\right), \; F_{lair} = L \times \left(\pi - |\theta_{pl}|\right)$$

and

$$F_{obstacles_j} = C_j / R^2_{ij}$$

that is a Coulomb's field generated by point sources where, R^2_{ij} is the distance between the predator *ith* and the obstacle *jth*, C_j are parameters that depend on directional range sensors. The term F_{clash} represents the clash between two predators. It has been represented with a short interaction field:

$$F_{clash} = \begin{cases} -C & if\, d \le d_1 \\ -C\dfrac{d-d_2}{d_2-d_2} & if\, d_1 \le d \le d_2 \\ 0 & if\, d > d_2 \end{cases}$$

the meaning of d_1 and d_2 are given in Fig. 5.

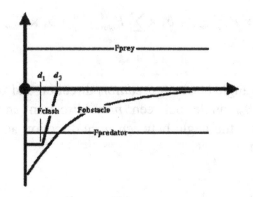

Fig.5. The fitting function

The memory model. The visual memory has been implemented by
a decreasing S-function:

$$S(p,a,b,c)=\begin{cases}1 & p\le a\\[2mm] 1-2\left(\dfrac{p\text{-}a}{c\text{-}a}\right)^{2} & a<p\le b\\[3mm] 2\left(\dfrac{p\text{-}c}{c\text{-}a}\right)^{2} & b<p\le c\\[2mm] 0 & p>c\end{cases}$$

with $b= (a+c)/2$. This choice reflects the fact that the prey and a
predator will forgive after a while the direction of the lair and the
prey respectively. The memory is refreshed any time the prey
(predator) sees the lair (prey).

4   Implementation and experimental result

A simulator of prey-predators system has been implemented in C++
in order to perform experiments. A visual interface has been
designed to easily introduce experimental conditions (number of
predators, their speed, position of prey, predators and lair).

Moreover, users may define the field of view of each agent and the characteristics of each detector (video and range) (see Fig. 6.).

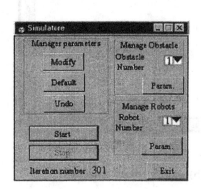 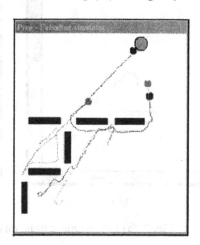

Fig. 6. The user interface

The GA has been designed to approximate the solution of the general optimal problem above stated. The planner, on the basis of the information given by the sensors agents, provides the direction of the displacement of each agent (prey, predators). The structure of the GA is that introduced before.

The starting population represents 200 directions chosen at random. The crossover is performed in a single point and the probability of crossover is uniform. The probability of bit-mutation has been set to $p=0.01$. The selection of the new population is performed by using the tournament technique.

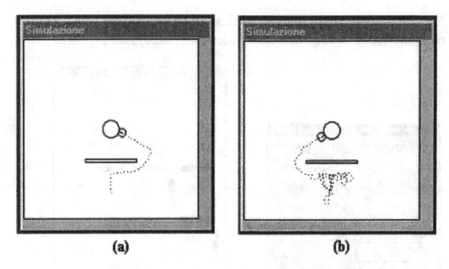

Fig. 7. (a) long memory trajectory; (b) short memory trajectory.

Experiments have been carried out to study the role of the memory in finding the solution. Fig. 7. shows the trajectory of the prey in the case of long memory $c=30$ and short memory $c=10$ without predators. In the first case the prey reaches the lair after 29 steps; in the second case after 96 steps.

Fig. 8. shows the mean number of successes of the prey for reaching the lair versus its memory length. The curve has been obtained by performing the experiments 100 times. Fig. 9. shows the mean number of steps, P, that are necessary to the predator for reaching the prey versus the predator memory length. Both experiments have been performed with the same initial conditions in the presence of 3 predators and one prey. Note that the initial speed and positions of the prey and predators is always the same in all trials.

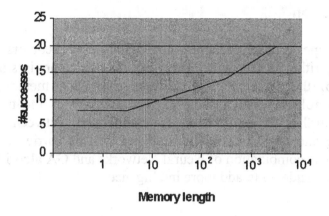

Fig. 8. Number of successes to reach the lair versus the memory length.

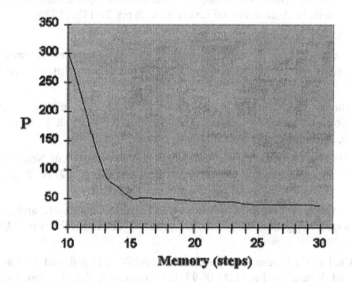

Fig. 9. Steps necessary to reach the prey versus its memory length.

The experiments in Fig. 8 and Fig. 9 show that the S-memory enhances the performance in discovering near-optimal trajectories. In fact, the S-memory includes the ability of building a temporary mental description of the environment, and this feature is common to most of living systems.

Conclusion

In this paper is described a new approach to prey predators problem. The algorithm is GA with S-memory. In particular, is studied the effect of the S-memory. The results show that memory play a relevant role in system survival. The stability of the trajectories needs further investigations. A comparison between different learning and problem solving strategies will be carried out in the future. The combination of Neural Networks and GA algorithms will be also considered to add more intelligence.

References

[1] V. Volterra, Sulle variazioni e fluttuazioni di specie animali conviventi, Memorie della R. Accademia dei Lincei S., 6, 2, pp. 31-113, (1926).

[2] A.J. Lotka, Proc. Nat. Accad. Sci., USA 6, 410, (1920).

[3] R. Levins, Some demographic and genetic consequences of environmental heterogeneity for biological control, Bull Entomol. Soc. Am 15, pp. 237-240, (1969).

[4] T. Haynes and S. Sen, Evolving behavioral strategies in predators and prey, in Adaptation and Learning in Multiagent Systems (G.Weißand S. Sen, eds.), pp. 113-126, Berlin: Springer Verlag, 1996.

[5] H. Hautop Lund, D. Parisi, Preadaptation in Populations of Neural Networks Evolving in a Changing Environment, in Artificial Life,Vol.2, N. 2, pp.179-197, (1995).

[6] V.Di Gesù, B.Lenzitti, G.Lo Bosco, D.Tegolo, A distributed architecture for autonomous navigation of robots, in proc. of IEEE conference CAMP2000, Padova, 2000.

[7] A.Chella, V.Di Gesù, S.Gaglio, et.al., "DAISY: a Distributed Architecture for Intelligent System", in Proc.CAMP-97, IEEE Computer Soc., Boston, 1997.

[8] V. Rai, V. Kumar, L.K. Pande, A New Prey-Predator Model, IEEE Transaction on Systems, Man, and Cybernetics, Vol.21, no.1, (1991).

[9] A. Yakubu, Pery Dominance in Discrete Predator-Prey Systems with a Prey Refuge, Mathematical Biosciences, vol. 144, pp.155-178, (1997).

[10] V. Di Gesù, Multi-agent System and Artificial Vision, in Computer and Devices for Communication, pp. 393-402, (1998).

[11] H. Muhlenbein and D. Schilierkamp-Voosen, Predictive Models for the Breeder Genetic Algorithm - Continuous Parameter Optimization, Evolutionary Computing, v1 n1, pp. 25-49, (1993).

Qualitative Models and Fuzzy Systems: An Integrated Approach to System Identification

Riccardo Bellazzi
Dipartimento di Informatica e Sistemistica – Università di Pavia, Pavia, Italy
Raffaella Guglielmann
Dipartimento di Matematica - Università di Pavia, Pavia, Italy
Liliana Ironi
Istituto di Analisi Numerica - C.N.R., Pavia, Italy

Abstract: We present a fuzzy-neuro method for the identification of nonlinear dynamical systems. The key idea which underlies our approach consists in the integration of qualitative modeling methods with fuzzy systems. The fuzzy model is initialized from rules which express the transition from one state to the next one. Such rules are automatically built by encoding the qualitative descriptions of the system dynamic behaviors inferred by the simulation of the qualitative model. The major advantage which derives from such an integrated framework lies in its capability both to represent the structural knowledge of the system at study and to determine, by exploiting the available experimental data, a functional approximation of the system dynamics that can be used as a reasonable predictor of the system's future state. Results obtained by the application of our method for identification of the intracellular kinetics of Thiamine from data collected in the intestine cells will be discussed.

1   Introduction

A suitable solution for the approximation of nonlinear dynamical systems is represented by the so-called fuzzy-neural approximators [9]. Such identifiers are known to hold the universal approximation property [9], and they are able to handle both experimental data and a priori knowledge on the unknown system dynamics, in the form of IF-THEN FUZZY RULES (FR) However, direct information, in the linguistic form, on the system dynamics is often poor or unavailable. Therefore, also the fuzzy-neural inference may become extremely inefficient since both the structure, i.e. the number of FR's, and its parameters have to be estimated from input-output data [4,9].

An alternative way to express the available, but incomplete knowledge about the system dynamics is represented by qualitative differential equations, where functional relationships between the problem variables are defined in terms of regions of monotonicity. Qualitative modeling approaches have been proposed [5] to cope with incomplete knowledge. Such approaches provide both formalisms for qualitative differential modeling and methods for qualitative simulation. Qualitative simulation allows us to derive all of the behavioral distinctions of the system under study from one of its initial state.

In this paper we propose a method, called FS-QM which combines the qualitative and fuzzy-neural modeling frameworks. It aims at solving the crucial problem of the construction of a meaningful FUZZY RULE BASE (FRB), and consequently of a fuzzy-neural system able to approximate the system dynamics. It is applicable whenever the incompleteness of the available physical knowledge of the system under study does not prevent from formulating a qualitative differential model of its dynamical behavior. The resulting hybrid method allows us to initialize properly both the fuzzy-neural network and the parameters with a consequent significant improvement in computational efficiency during the training phase. The embedment of all the available structural knowledge into the fuzzy identifier allows us to find a robust model in those cases that are intractable with standard modeling techniques. A great deal of application domains may take advantage of our approach: among those, medicine seems to be a natural application field as prior structural knowledge is generally available but incomplete, and experimental data set are not rich enough for a successful exploitation of fuzzy-neural methods, namely for the extraction of a meaningful fuzzy rule base.

2   Method

We deal with the general problem of approximating discrete nonlinear systems, described by the following input-output equation:

$$y_{k+1} = f(\underline{u}_k, y_k, \underline{\theta}) \tag{1}$$

where $\underline{u} \in \Re^{n-1}$ and $y \in \Re$ are discrete-time sequences, $\underline{\theta}$ is the vector of parameters, and the function $f(\cdot)$ is in general unknown. Our goal is therefore to find a continuous function approximator

$$f(\underline{x}, \underline{\theta}) = \frac{\sum_{j=1}^{M} \hat{y}_j \prod_{i=1}^{n} \exp(-(\frac{x_i - \hat{x}_i^j}{\sigma_i^j})^2)}{\sum_{j=1}^{M} \prod_{i=1}^{n} \exp(-(\frac{x_i - \hat{x}_i^j}{\sigma_i^j})^2)} \tag{2}$$

$f(\underline{x}_k, \underline{\theta})$ where $\underline{x}_k = \{\underline{u}_k, y_k\}$ $\forall k$.

In our work we use a Fuzzy System (FS) to build $f(\underline{x}_k, \underline{\theta})$: by exploiting FS's with the *singleton fuzzifier*, the *product inference rule*, the *center average defuzzifier*, and under the assumption of *Gaussian membership functions*, we obtain the following expression [9]:

where M is the number of FR's, and $\hat{y}_j$ is the point where the membership function of the output in the j-th rule reaches its maximum value.

The parameters $\underline{\theta} = \{\hat{\underline{y}}, \hat{\underline{x}}, \underline{\sigma}\}$ where $\hat{\underline{y}} = \{\hat{y}_j\}, \hat{\underline{x}} = \{\hat{x}_i^j\}, \underline{\sigma} = \{\sigma_i^j\}$, and $i = 1, \cdots, n$, $j = 1, \cdots, M$, are usually identified from a set of experimental data.

The approximator derived in equation (2) is known to possess some desirable properties for several classes of membership functions, like the capability of approximating any continuous function with an arbitrary degree of accuracy [9]. Moreover, the proposed FS can be represented through a three-layer neural network, and therefore trained by using a Back-Propagation (BP) scheme [9]. The capability of FS's to express the a priori knowledge on the domain could be fully exploited if we were able to (i) define the number of rules necessary to describe the dynamics of the system, (ii) provide for a good initialization of the back-propagation algorithm, through a choice of the initial values for $\underline{\theta}$. Unfortunately, in the majority of published papers, the FS is directly inferred from data, and no formal methods have been proposed to derive it and express it from the knowledge available on the structure of the system. In order to cope with this problem, we propose a new method for FS initialization based on qualitative simulation. Figure 1 sketches the main steps of the method.

Qualitative simulation (QSIM, [5]) derives qualitative descriptions of the possible behaviors of a dynamical system from a qualitative representation both of its physical structure and of an initial state $QS(t_0)$. The system is represented through a Qualitative Differential Equation (QDE), that describes a system in the same terms as an ordinary differential equation does, except that (i) the values of variables are qualitatively expressed in terms of their ordinal relations with landmark values, and (ii) functional relationships between variables are described in terms of regions of monotonicity. The set of behaviors $\{B_1, \cdots, B_m\}$ generated by QSIM includes all possible behaviors of the system described by the qualitative differential equation and the initial state. Such a set is finite due to the qualitative level of description, and it is represented by a tree. Each behavior is a finite sequence of qualitative states, that represents a possible temporal evolution of the system. The qualitative values are represented through landmark values: the real values the variables assume are mapped into a finite ordered set Q_L, called *qualitative quantity-space*, whose elements are landmark values, l_k , and open intervals, (l_{k-1}, l_k), bounded by two adjacent landmark values. A landmark value is a symbolic name for a particular real number, whose value may be unknown, and defines regions where qualitative system properties hold. The landmark-based representation allows us to express incomplete knowledge about values of variables as they are defined only by their order relations with the elements of Q_L.

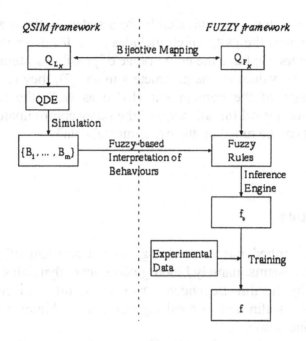

Fig. 1. Main steps of the method

Fuzzy sets may be used to represent qualitative regions, too: the underlying range of a real variable can be discretized into a finite ordered set Q_F, that we call *fuzzy quantity-space*, whose elements are fuzzy sets. In this representation, the qualitative regions are expressed through membership functions, which may be viewed as a measure of the suitability of applying a given qualitative description to a state variable.

Whenever the domain knowledge allows us to define the sets Q_L and Q_F, we can establish a one-to-one correspondence between their elements: in such a way a real value can be represented in both frameworks. Then, on the basis of this mapping, we can translate the tree of behaviors into the fuzzy formalism, i.e. we build the fuzzy rules from the qualitative behaviors generated by QSIM: the resulting fuzzy rules can be seen as a measure of the possible transition from qualitative regions, or equivalently from states, to the next ones. As a consequence, the FRB, which includes the rules generated from all the behaviors, captures the entire range of possible system dynamics. As far as the values of the parameters in eq. (2), they are initialized on the basis of the domain knowledge as well: the resulting FS provides for a good initialization of the fuzzy approximator searched for, and it can be tuned on the experimental data.

3   Results

We have applied our methodology to the problem of identifying metabolic systems, namely blood glucose level dynamics in patients affected by Insulin--Dependent Diabetes Mellitus in response to exogenous insulin and to meal ingestion, and Thiamine kinetics in the intestine tissue.

In this paper let us focus on the second system. Thiamine (Th), also known as vitamin B_1, is one of the basic micronutrients present in food and essential for health. Within the cells, Th participates in the carbohydrate metabolism, in the central and peripheral nerve cell function and in the myocardial function. Many studies based on the compartmental modeling approach have been carried out to quantitatively assess the intracellular Thiamine kinetics in several tissues, both in normal and pathological conditions [6-8]. From the modeling viewpoint these studies exploit linear differential equations, as they consider the system around the steady state condition: unfortunately the linearity hypothesis revealed physiologically unsound. On the other hand, the lack of structural knowledge prevented from formulating a nonlinear compartmental model. Moreover, the small number of the available samples has also prevented from a successful representation of the system dynamics through classical input-output approaches. For such reasons, we decided to apply our method with the aim to build a nonlinear model of this system.

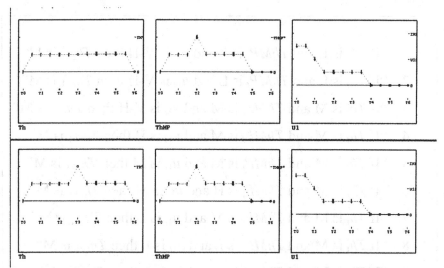

Fig. 2. Qualitative behaviors obtained by simulating the model of the chemical form Th. On each row the plots describe the dynamics of the model's variables.

In particular, the structure of the compartmental model that describes thiamine intracellular behavior has been used to derive three QDE's, each one representing the dynamics of the chemical form in which thiamine is present in cells, namely thiamine (Th), thiamine piro-phosphate (ThPP) and thiamine mono-phosphate (ThMP). Each qualitative model has been simulated using QSIM, and a fuzzy approximator as in eq. (2) has been derived for each chemical form by exploiting the rules extracted from the simulated qualitative behaviors. In the following we will fix our attention on the first chemical form: Figure 2 and Table 1 show the qualitative behaviors and their fuzzy interpretation into rules, respectively; the generated FRB (see Table 1) allows us to derive the structure of the fuzzy approximator.

1. "If Th_t is L and $ThMP_t$ is L and u_{1t} is V-H then Th_{t+1} is L"

2. "If Th_t is L and $ThMP_t$ is L and u_{1t} is V-H then Th_{t+1} is M"

3. "If Th_t is M and $ThMP_t$ is M and u_{1t} is V-H then Th_{t+1} is M"

4. "If Th_t is M and $ThMP_t$ is M and u_{1t} is H then Th_{t+1} is M"

5. "If Th_t is M and $ThMP_t$ is M and u_{1t} is M then Th_{t+1} is M"

6. "If Th_t is M and $ThMP_t$ is H and u_{1t} is M then Th_{t+1} is M"

7. "If Th_t is M and $ThMP_t$ is M and u_{1t} is L then Th_{t+1} is M"

8. "If Th_t is M and $ThMP_t$ is L and u_{1t} is L then Th_{t+1} is M"

9. "If Th_t is M and $ThMP_t$ is L and u_{1t} is L then Th_{t+1} is L"

10. "If Th_t is M and $ThMP_t$ is M and u_{1t} is M then Th_{t+1} is H"

11. "If Th_t is H and $ThMP_t$ is M and u_{1t} is M then Th_{t+1} is M"

Table 1. The FRB derived from the simulated behaviors in Fig. 2.

Then, we have identified the fuzzy models on experimental data obtained with tracer experiments performed on rats. The identified models have been validated on a new data set. The accuracy of the whole model has been tested on a new data set in accordance with a *parallel scheme* where only the current inputs to the overall system are measured data, whereas the current output and input to each subsystem are simulated values. At each step, the performance has been compared with fuzzy-neural identifiers obtained following a data-driven (FS-DD) approach: the vector of parameters has been initialized from the numerical evidence according to the procedure described in [9], and then tuned on the experimental data. The structures, i.e. the number of rules M, of the identifiers have been determined on the basis of a well-known model selection index, called Akaike Information Criterion (AIC), defined as follows:

$$AIC = N \log(V) + 2 p$$

where N is the number of available data, p is the number of parameters (depending on M), and V is the loss function, namely the sum of squared errors, in correspondence with the optimal estimate of θ. The AIC is calculated for increasing values of p: the "best" model structure corresponds to the value of p which minimizes the AIC index.

3.1 Identification

Although the problem is ill-posed due to the small number of data, our proposed method performs quite well: even though the number of rules generated for the Th subsytem $(M=11)$ is much higher than that one of the corresponding optimal FS-DD identifier $(M=6)$, the required accuracy (10^{-4}) is achieved in a very low number of BP loops (see Table 2); this result can be explained by the goodness of the initialization of both the identifier structure and the guess of parameters. On the contrary, the FS-DD approach, initialized straightforward from the data, does not reach the desired error threshold as it gets trapped in a local minimum.

Chemical form	No. of BP-loops		Identification error		Validation error	
	FS-DD	FS-QM	FS-DD	FS-QM	FS-DD	FS-QM
Th	523	64	$1.295\cdot10^{-4}$	$9.862\cdot10^{-5}$	$1.736\cdot10^{-2}$	$4.6\ 10^{-3}$
ThPP	308	32	$1.619\cdot10^{-4}$	$9.376\cdot10^{-5}$	$1.61\cdot10^{-2}$	$1.15\cdot10^{-2}$
ThMP	172	32	$1.431\cdot10^{-4}$	$7.472\cdot10^{-5}$	$2.574\cdot10^{-3}$	$2.41\cdot10^{-3}$

Table 2. Comparison of performance of FS-QM approach with FS-DD. Both identification and validation errors are calculated as the mean squared error.

3.2 Validation and simulation

Each identified approximator has been validated on a different set of data collected in an independent experiment, still on normal subjects. The results confirm the robustness of our approach. Figure 3 highlights the failure of the FS-DD approach to reproduce the experimental profile.

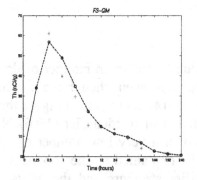 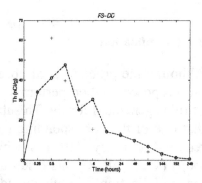

Fig. 3. The subsystem: validation results of the model identified with both approaches. The model predictions (°) are compared with actual data (+).

Our final goal is the construction of a simulator of the overall system dynamics that is capable to reproduce the system behavior in response to any input signals, at least in the range of the experimental settings previously defined. Figure 4 shows the simulation results on the validation data set: we can observe a really good fit between the data and the predictions provided by our method, whereas the FS-DD identifiers are not robust enough to simulate the overall system dynamics. The simulation results clearly show the validity of the proposed approach as an alternative methodology to identify nonlinear systems in presence of incomplete structural knowledge.

4 Further work

Our proposed method seems to be potentially exploitable in a variety of domains and application problems. Thanks to its capability of automatically encoding prior knowledge, it allows us to use fuzzy-neuro systems in data poor contexts, such as metabolic modeling, as well as in data rich environment, such as closed loop control in Diabetes or hemodialysis. From the methodological point of view, a

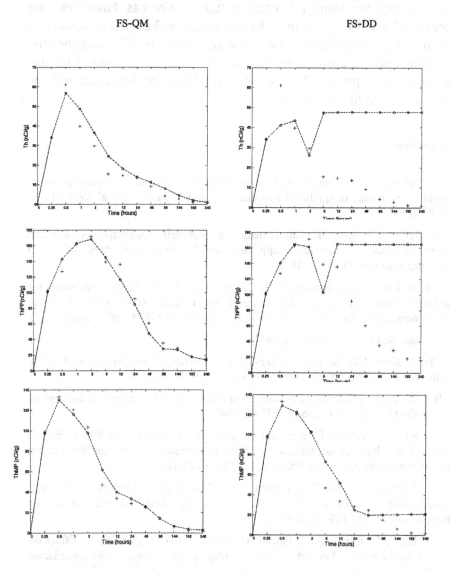

Fig. 4. Simulation results of the overall system obtained with with both FS-QM and FS-DD identifiers. The simulated predictions (°) are compared with actual data (+).

crucial step in the definition of the method is how to represent properly landmarks and intervals between them in the fuzzy framework. At the current state of the implementation, Gaussian membership functions are used: actually, whereas landmarks are suitably described by them, intervals could be better represented by means of bell-shaped or "pseudo-trapezoidal" membership functions, which assume the maximum value in an interval instead of in a single point. The use of these kinds of functions will be investigated in the near future.

References

1. R. Bellazzi, R. Guglielmann, L. Ironi. How to improve fuzzy-neural system modeling by means of qualitative simulation, *IEEE Trans. Neural Networks* 11 (2000) 249-254.

2. R. Bellazzi, L. Ironi, R. Guglielmann, M. Stefanelli. Qualitative models and Fuzzy Systems: an integrated approach for learning from data, *Artificial Intelligence in Medicine* 14 (1998) 5-28.

3. R. Bellazzi, R. Guglielmann, L. Ironi,. A qualitative-fuzzy framework for nonlinear black-box system identification, in: T. Dean (ed.), *Proc. Sixteenth International Joint Conference on Artificial Intelligence (IJCAI 99)* (Morgan

Kaufmann, San Francisco, 1999), 1041-1046.

4. H.M. Kim, J.M. Mendel, Fuzzy Basis Functions: Comparison with Other Basis Functions, *IEEE Trans. Fuzzy Systems*, 3 (1995) 158-168.

5. B. J. Kuipers. *Qualitative Reasoning: modeling and simulation with incomplete knowledge* (MIT Press, Cambridge MA, 1994).

6. A. Nauti, C. Patrini, C. Reggiani, A. Merighi, R. Bellazzi, G. Rindi, In vivo study of the kinetics of thiamine and its phosphates in the deafferented rat cerebellum, *Brain Metabolic Disease*, 12 (1997) 145-160.

7. G. Rindi, C. Patrini, V. Comincioli and C. Reggiani, Thiamine content and turnover rates of some rats nervous regions, using labeled Thiamine as tracer, *Brain Research*, 181 (1980) 369-380

8. G. Rindi, C. Reggiani, C. Patrini, G. Gastaldi and U. Laforenza, Effect on ethanol on the in vivo kinetics of thiamine phosphorilation and dephosphorilation in different organs-II, *Acute effects Alcohol and Alcoholism*, 27 (1992) 505-522.

9. L.X. Wang, *Adaptive Fuzzy Systems and Control* (Prentice Hall, Engelwood Cliffs, N.J., 1994).

Fuzzy Reliability Analysis of Concrete Structures by Using a Genetically Powered Simulation

Fabio Biondini[1], Franco Bontempi[2], and Pier Giorgio Malerba[3]

[1] Department of Structural Engineering, Technical University of Milan
P.za L. da Vinci, 32 – 20133 Milan, Italy
biondini@stru.polimi.it

[2] Department of Structural and Geotechnical Engineering, University of Rome "La Sapienza"
Via Eudossiana, 18 – 00184 Roma, Italy
franco.bontempi@uniroma1.it

[3] Department of Civil Engineering, University of Udine
Via delle Scienze, 208 - 33100 Udine
piergiorgio.malerba@dic.uniud.it

Abstract. The paper deals with the reliability assessment of concrete framed structures. Due to the uncertainties involved in the problem, the geometrical and mechanical properties which define the structural problem cannot be considered as deterministic quantities. In this work, such uncertainties are modeled by using a fuzzy criterion which considers the model parameters bounded between minimum and maximum suitable values. The problem is formulated in terms of safety factor and the membership function over the failure interval is defined for several limit states by using a simulation technique. In particular, the strategic planning of the simulation is here found by a genetic optimization algorithm and the structural analyses are carried out by taking both material and geometrical non-linearity into account. An application to a prestressed concrete continuous beam shows the effectiveness of the proposed procedure.

1 Introduction

As known, concrete structures exhibit a structural behavior affected by several sources of non-linearity: the constitutive laws of the materials (concrete, reinforcing and prestressing steels) and the geometrical effects induced by the change of the configuration. Consequently, the reliability of a structure belonging to this class of systems cannot be definitely assured without considering its actual non-linear behavior. In this context, thought the reliability of the structure as resulting from a general and comprehensive investigation of all its failure modes, one must pay attention to the following aspects of the assessment process:

– *Available Data and Sources of Uncertainty.*
– *Limit States of Failure and Safety Factor.*
– *Structural Model and Non-Linear Analysis.*
– *Simulation Process and Synthesis of the Results.*

2 Available Data and Sources of Uncertainty

The methodology presented in the following has general validity and is here directly applied to the reliability analysis of the two-span post-tensioned reinforced concrete continuous beam shown in Figure 1. The span length is $l = 7500\,\text{mm}$, while the dimensions of the rectangular cross-section are $b = 203.2\,\text{mm}$ and $h = 406.4\,\text{mm}$. The beam is reinforced with 2 bars $\varnothing 14\,\text{mm}$ ($A_{s1} = 153.9\,\text{mm}^2$) placed both at the top and at the bottom edge with a cover $c = 32.4\,\text{mm}$. The prestressing steel tendon consists of 32 wires $\varnothing 5\,\text{mm}$ ($A_{p1} = 19.6\,\text{mm}^2$) adherent to the concrete, having a straight profile from the ends of the beam ($e_1 = 0$) to the middle of its spans ($e_{2nom} = -50\,\text{mm}$) and a parabolic profile from these points to the middle support ($e_3 = 88\,\text{mm}$). After the time-dependent losses, at the ends the nominal prestressing force is $P_{nom} = 527.5\,\text{kN}$. This force decreases along the beam because of the losses due to friction. The curvature friction coefficient $\mu = 0.3$ and the wobble friction coefficient $K = 0.0016\,\text{rad/m}$ have been assumed. The material properties are defined by the following nominal values:

– concrete: $\qquad f_{c,nom} = -41.3\,\text{MPa}$ $\quad \varepsilon_{c1} = -2\text{‰}$ $\quad \varepsilon_{cu} = -3.4\text{‰}$ $\quad \varepsilon_{ctu} = 2\varepsilon_{ct1}$

– reinforcing steel: $f_{sy,nom} = 314\,\text{MPa}$ $\quad E_s = 196\,\text{GPa}$ $\quad \varepsilon_{su} = 16\%$

– prestressing steel: $f_{py,nom} = 1480\,\text{MPa}$ $\quad E_p = 200\,\text{GPa}$ $\qquad \varepsilon_{pu} = 1\%$

with a weight density of the composite material $\gamma = 25\,\text{kN/m}^3$. Finally, two concentrated live loads Q placed at the distance $a = 4880\,\text{mm}$ from each end as shown in Figure 1 have been considered.

To the aim of the reliability analysis, the following 32 quantities which define the structural system are considered to be uncertain:

- the strengths of concrete and of both reinforcing and prestressing steel in each of ten segments in which the beam is assumed to be subdivided (30 variables);
- the prestressing force of the strands (1 variable);
- the live loads (1 variable).

Such 32 quantities $\mathbf{x} = [x_1\ x_2\ ...\ x_{32}]^T$ are modeled as fuzzy variables having a triangular membership function with interval base [0.70–1.30] and mean value equal 1.00. Seven α-levels are considered, corresponding to the following intervals: [0.70–1.30], [0.75–1.25], [0.80–1.20], [0.85–1.15], [0.90–1.10], [0.95–1.05], and [1.00–1.00].

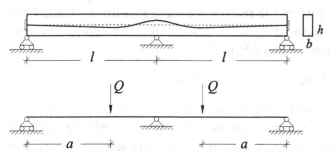

Fig. 1. Prestressed continuous beam (Lin 1955). Tendon layout and load condition.

3 Limit States of Failure and Safety Factor

Based on the general concepts of the reinforced concrete (R.C.) and prestressed concrete (P.C.) design, the structural performances should generally be described with reference to a specified set of limit states, as regards both serviceability and ultimate conditions, which separate desired states of the structure from undesired ones (Bontempi *et al.* 1998). Splitting cracks and considerable creep effects may occur if the compression stresses σ_c in concrete are too high. Besides, excessive stresses either in reinforcing steel σ_s or in prestressing steel σ_p can lead to unacceptable crack patterns. Excessive displacements s may also involve loss of serviceability and then have to be limited within assigned bounds s^- and s^+. Based on these considerations, the following limitations account for adequate durability at the serviceability stage (*Serviceability Limit States*):

1. $-\sigma_c \le -\alpha_c f_c$ **2.** $|\sigma_s| \le \alpha_s f_{sy}$ **3.** $|\sigma_p| \le \alpha_p f_{py}$ **4.** $s^- \le s \le s^+$ (2)

where α_c, α_s and α_p are suitable reduction factors of the strengths f_c, f_{sy} and f_{py}. In particular, for the prestressed beam the serviceability limit states are detected by assuming $\alpha_c = 0.45$, $\alpha_s = 0.80$, $\alpha_p = 0.75$, $s^+ = -s^- = 1/400$.

When the strain in concrete ε_c, or in the reinforcing steel ε_s, or in the prestressing steel ε_p reaches a limit ε_{cu}, ε_{su} or ε_{pu}, respectively, the collapse of the corresponding cross-section occurs. However, the collapse of a single cross-section doesn't necessarily lead to the collapse of the whole structure, the latter is caused by the loss of equilibrium arising when the reactions **r** requested for the loads **f** can no longer be developed. So, the following ultimate conditions have to be verified (*Ultimate Limit States*):

1. $-\varepsilon_c \le -\varepsilon_{cu}$ **2.** $|\varepsilon_s| \le \varepsilon_{su}$ **3.** $|\varepsilon_p| \le \varepsilon_{pu}$ **4.** $\mathbf{f} \le \mathbf{r}$ (3)

Since these limit states refer to internal quantities of the system, a check of the structural performance through a nonlinear analysis needs to be carried out at the load level. To this aim, it is useful to assume $\mathbf{f} = \mathbf{g} + \lambda\mathbf{q}$, where **g** is a vector of dead loads and **q** a vector of live loads whose intensity varies proportionally to a unique multiplier $\lambda \ge 0$.

4   Structural Model and Non-linear Analysis

In most cases, R.C. and P.C. structures should be analyzed by taking material and, eventually, geometrical non-linearity into account if realistic results under all load levels are needed. In this work, a two-dimensional structure is modeled using a R.C./P.C. beam finite element whose formulation, based on the Bernoulli-Navier hypothesis, deals with such kinds of non-linearity (Bontempi *et al.* 1995, Malerba 1998).

In particular, both material $\mathbf{K}'_M$ and geometrical $\mathbf{K}'_G$ contributes to the element stiffness matrix $\mathbf{K}'$ and the nodal forces vector $\mathbf{f}'$, equivalent to the applied loads $\mathbf{f}'_0$ and to the prestressing $\mathbf{f}'_p$, are derived by applying the principle of the virtual displacements and then evaluated by numerical integration over the length l of the beam:

$$K' = K'_M + K'_G \qquad K'_M = \int_0^l B^T H B \, dx \qquad K'_G = \int_0^l N \, G^T G \, dx \qquad f' = \int_0^l N^T (f'_0 + f'_p) \, dx \quad (4)$$

$$N = \left[\begin{array}{c|c} N_a & 0 \\ \hline 0 & N_b \end{array} \right] \qquad B = \left[\begin{array}{c|c} \partial N_a / \partial x & 0 \\ \hline 0 & \partial^2 N_b / \partial x^2 \end{array} \right] \qquad G = \left[\begin{array}{c|c} 0 & \dfrac{\partial N_b}{\partial x} \end{array} \right] \quad (5)$$

where N is the axial force and $\mathbf{N}$ is a matrix of axial N_a and bending N_b displacement functions. In the following, the shape functions of a linear elastic beam element having uniform cross-sectional stiffness $\mathbf{H}$ and loaded only at its ends are adopted. However, due to material non-linearity, the cross-sectional stiffness distribution along the beam is non uniform even for prismatic members with uniform reinforcement. Thus, the matrix $\mathbf{H}$, as well as the sectional load vector equivalent to the prestressing f'_p, have to be computed for each section by integration over the area of the composite element, or by assembling the contributes of concrete and steel. In particular, the contribute of the concrete is evaluated by subdividing the area of the beam cross section in four-nodes isoparametric sub-domains and by performing a Gauss-Legendre and/or a Gauss-Lobatto numerical integration over each sub-domain and then along the whole element. In this way, after the constitutive laws of the materials are specified, the matrix $\mathbf{H}$ of each section can be computed under all load levels.

The equilibrium conditions of the beam element are derived from the already mentioned principle of the virtual work. Thus, by assembling the stiffness matrix $\mathbf{K}$ and the vectors of the nodal forces $\mathbf{f}$ with reference to a global coordinate system, the equilibrium of the whole structure can be formally expressed as follows:

$$\mathbf{K} \, \mathbf{s} = \mathbf{f} \qquad (6)$$

where $\mathbf{s}$ is the vector of the nodal displacements. It is worth noting that the vectors $\mathbf{f}$ and $\mathbf{s}$ have to be considered as total or incremental quantities depending on the nature of the stiffness matrix $\mathbf{K} = \mathbf{K(s)}$, or if a secant or a tangent formulation is adopted.

In the considered application the stress-strain diagram of concrete is described by the Saenz's law in compression and by an elastic perfectly-plastic model in tension. The stress-strain diagram of reinforcing steel is assumed elastic perfectly-plastic in tension and in compression. For prestressing steel the plastic branch is instead assumed as nonlinear and described by a fifth degree polynomial function (Bontempi *et al.* 1998).

5   Simulation Process and Synthesis of the Results

Let p a parameter belonging to the set of quantities which define the structural problem. It is clear that for each structural state, i.e., each parameter set value, one load multiplier λ value can be determined for each limit state, i.e., a set of limit load values can be determined. For the sake of simplicity, we start our developments by considering the relationship between one single parameter p and one single limit state, defined by its corresponding limit load multiplier λ. At first, it is worth noting that, in general, such relationship is nonlinear even if the behavior of the system is linear. This is typical of the design processes where the structural properties which correlate loads and displacements are considered as design variables. Thus, the nonlinear law $\lambda=\lambda(p)$ can be drawn as in Figure 2.a, which shows that for each value of p, there is a corresponding value of λ. However, from Figure 2.b it is also clear that the response interval

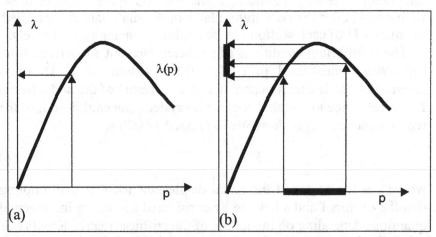

Figure 2. (a) Relationship between a structural parameter p and a limit load multiplier λ. (b) Interval of the limit multiplier λ corresponding to an interval of the parameter p.

$[\lambda_{min}; \lambda_{max}]$ corresponding to $[p_{min}; p_{max}]$ cannot be simply obtained from $\lambda(p_{min})$ and $\lambda(p_{max})$. The problem of finding the interval response can be instead properly formulated as an optimization problem by assuming the objective function to be maximized as the size of the response interval itself. In particular, for the general case of n independent parameter p, collected in a vector $\mathbf{x} = [p_1 \ p_2 \ ... \ p_n]^T$, and m assigned limit states, the following objective function is introduced:

$$F(\mathbf{x}) = \sum_{i=1}^{m} \left(\lambda_{i,\max} - \lambda_{i,\min} \right) \tag{7}$$

A solution $\mathbf{x}$ of the optimization problem which takes the side constraints $\mathbf{x}_{\min} \leq \mathbf{x} \leq \mathbf{x}_{\max}$ into account is developed by genetic algorithms. Such algorithms are heuristic search techniques which belong to the class of stochastic algorithms, since they combine elements of deterministic and probabilistic search (Michalewicz 1992).

Properly, the search strategy works on a *population* of *individuals* subjected to an evolutionary process, where individuals compete between them to survive in proportion to their *fitness* with the *environment*. In this process, population undergoes continuous reproduction by means of some *genetic operators* which, because of competition, tend to preserve best individuals. From this evolutionary mechanism, two conflicting trends appear: exploiting of the best individuals and exploring the environment. Thus, the effectiveness of the genetic search depends on a balance between them, or between two principal properties of the system, *population diversity* and *selective pressure*. These aspects are in fact strongly related, since an increase in the selective pressure decreases the diversity of the population, and vice versa (Biondini 1999).

With reference to the optimization problem previously formulated, a population of m individuals belonging to the environment $E = \{ \mathbf{x} \mid \mathbf{x}^- \leq \mathbf{x} \leq \mathbf{x}^+ \}$ represents a collection $X = \{ \mathbf{x}_1 \ \mathbf{x}_2 \ ... \ \mathbf{x}_m \}$ of m possible solutions $\mathbf{x}_k^T = [x_1^k \ x_2^k \ ... \ x_n^k] \in E$, each defined by a set of n design variables x_i^k ($k = 1, ..., m$). To assure an appropriate hierarchical arrangement of the individuals, their fitness $F(\mathbf{x}) \geq 0$, which increases with the adaptability of $\mathbf{x}$ to its environment E, should be properly scaled. More details about the adopted scaling rules, internal coded representation of the population, genetic operators and termination criteria, can be found in a previous paper (Biondini 1999).

Based on the models so introduced, a sample consisting of about 5000 simulations has been carried out for the considered α-levels. The membership functions of the limit load multiplier for the limit states previously defined are presented in Figure 3. At first we observe that, for the chosen load value ($\lambda=1$), the limit states are not violated and the structure is safe with respect to the assumed α-levels. For increasing load values ($\lambda>1$) the third service limit state (about the stress in the prestressing steel) never appears, while for the other limit states it can be appreciated the spread of the uncertainty, especially for the larger α-levels. To this regards, it is worth noting that uncertainties larger than 15% seems to appear critical, in particular for the ultimate limit states. Figure 4 shows the histograms of frequency of the load multiplier λ resulting from about 400 simulations performed for a given α-level and by assuming alternatively: (a) a purely random choice of the data, and (b) a genetically driven simulation. A comparison of the results leads us to appreciate the higher capability of the genetic search in exploring the regions of the response interval where the limit state violations tends to occur.

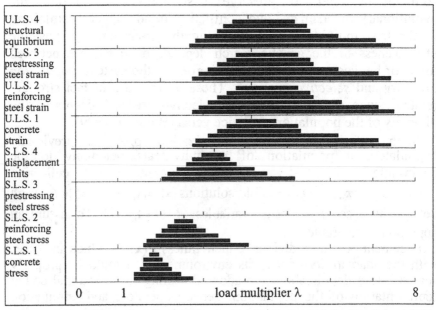

Figure 3. Resulting α-levels of the load multiplier λ for each limit state.

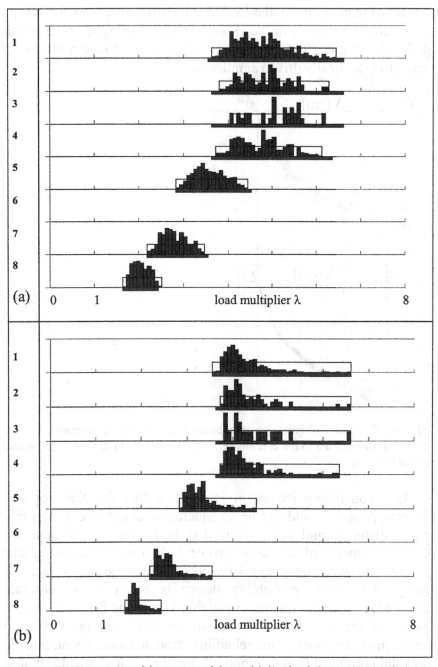

Figure 4. Histograms of frequency of the multiplier λ of about 400 simulations.

104

Finally, Figure 5 shows the load–displacement diagrams for several simulations associated to the α-levels corresponding to the intervals $[0.70 - 1.30]$ and $[0.95 - 1.05]$, respectively (a) Random choice of data. (b) Genetically driven simulation.

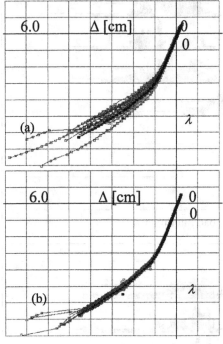

Figure 5. Load multiplier λ – maximum displacement Δ diagrams for 400 simulations associated to the α-levels (a) $[0.70–1.30]$ and (b) $[0.95–1.05]$ (nominal curves in black).

As a concluding remark, it is worth noting that the proposed fuzzy approach should not be considered as alternative to a purely probabilistic formulation, given that both methods account for different aspects of the same problem. In fact, as known, fuzzy theory allows a treatment of uncertainties due to a lack of information, while probability theory is based on an absolute knowledge about the stochastic variability resulting from the random nature of the same quantities. However, it is also noted that an autonomous approach to the reliability structural assessment, like the probabilistic formulation proposed by the codes, should find a higher rationality in a fuzzy approach which, due either to the real nature of the involved uncertainties, or to a higher simplicity of the mathematical formulation, seems to be more suitable for the design purposes.

References

1. Biondini F. Optimal Limit State Design of Concrete Structures using Genetic Algorithms. *Studi e Ricerche*, Scuola di Specializzazione in Costruzioni in Cemento Armato, Politecnico di Milano, **20**, (1999) 1-30.
2. Bojadziev G., M. Bojadziev. *Fuzzy sets, fuzzy logic, applications.* World Scientific (1995).
3. Bontempi F., P.G. Malerba and L. Romano. A Direct Secant Formulation for the Reinforced and Prestressed Concrete Frames Analysis. *Studi e Ricerche*, Scuola di Specializzazione in Costruzioni in Cemento Armato, Politecnico di Milano, **16**, (1995) 351-386 (in Italian).
4. Bontempi F., F. Biondini, and P.G. Malerba. Reliability Analysis of Reinforced Concrete Structures Based on a Monte Carlo Simulation. *Stochastic Structural Dynamics*. Spencer, B.F. Jr, Johnson E.A. (Eds.), Rotterdam: Balkema, 413-420 (1998).
5. Jang J.S.R., C.T. Sun, E. Mizutani. *Neuro-Fuzzy and Soft Computing.* Matlab Curriculum Series, Prentice Hall (1997).
6. Lin T.Y. Strength of Continuous Prestressed Concrete Beams Under Static and Repeated Loads. *ACI Journal*, **26**(10), (1955) 1037-1059.
7. Malerba P.G. (Ed.). *Limit and Nonlinear Analysis of Reinforced Concrete Structures.* Udine: CISM (1998) (in Italian).
8. Michalewicz Z. *Genetic Algorithms + Data Structures = Evolution Programs.* Berlin: Springer (1992).

System for Remote Diagnosis Based on Fuzzy Inference

Paolo Falcioni, Francesca Meloni, Irene Orienti

WRAP SpA
Viale Aristide Merloni, 47, 60044 Fabriano (AN)
paolo.falcioni@wraphome.com, francesca.meloni@wraphome.com,
irene.orienti@wraphome.com

Abstract: Object of this paper is the development of a non invasive diagnostic system applicable to any appliance. Specifically it is a remote monitoring system, based on local monitoring devices and a centralized expert system which receives the data from remote and processes them to obtain high level diagnostic information. The expert system has an inferential engine based on fuzzy computational methods.

1 Introduction

Aim of the project is to develop a system for remote diagnosis of appliances within user homes. A fundamental requirement is that the monitoring device is non invasive, so that it can be applied to already existing appliances, its cost must be adequate for the consumer market. These requirements have been satisfied by developing a processing system that can extract information from the electrical absorption of the appliance.

The developed system is composed by local measuring devices called "WESA" (WRAP Enabled Smart Adapter), which are inserted between the power socket and the appliance plug. The WESAs send electrical absorption patterns, via power line, to a telephone modem in the house and, from there, to a centralized expert system WESArd (WESA remote diagnostics). The expert system analyzes the information on the electrical absorption of the appliance and performs two fundamental diagnostic functions: it activates the fuzzy inference engine for detecting faults from subtle or slow changes, and it programs, from remote, the local devices enabling them to detect faults from abrupt, higher order changes.

2 Electrical absorption analysis

The inputs to the fuzzy inference engine are obtained by processing the data acquired by the local devices (the WESAs). The WESA is connected to the appliance socket and it measures its absorbed current. Experimental tests show that from current absorption patterns it is possible to:

- Detect single load activation
- Monitor each load current consumption

Moreover from a load current consumption pattern it is possible to:

- Detect a fault
- Determine the wear status of the load

2.1 Electrical loads detection and analysis

Detection of eletrical loads is based upon features extraction and classification. Each load is described by a set of features such as power absorption, power factor, average on time, etc...

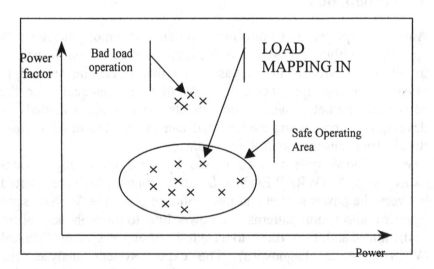

Fig. 1. A representation of loads in the feature space.

Such features form a distinctive cluster for each load in the features space [3][4].

The capability to detect a faulty load is based upon the following steps of operation:

1. Identifying the load cluster.
2. Recognize the load based on the cluster position.
3. Learning the load region in the feature space and establish maximum and minimum values.
4. Detect variation of load position in the feature space from its original position.
5. Use the variation as an input for the diagnostic inferential engine.

Steps 1 to 3 belong to the learning phase, while step 4 and 5 relate to the diagnostic inferential engine. In figure 1 is an ideal representation of the feature space. While figure 2 shows real loads mapping in the feature space from a refrigerator.

Fig. 2. Real loads mapping in the feature space from a refrigerator.

In the learning phase, the WESArd receives the data represented in figure 2, it classifies them in clusters and memorizes clusters regions for each load; in the monitoring phase input data are compared with learnt cluster region for load working validation.

When a load is not working properly its electrical features move slightly from the usual ones. Depending on the combination of electric features which have changed (power, power factor, energy, duty cycle, etc....) it is possible to recognize which fault has occurred out of a known set. The amount of change gives information on the degree to which the fault is present.

3. The diagnostic application

3.1 System architecture

Before describing the expert system, the overall system architecture is briefly explained. The local devices, WESAs, measure the appliance power absorption and send associated compressed data to a server hosting the WESArd expert system. Power line carrier technology is used within a home to convey data from one or more WESAs to a specific modem. WRAP technology provides for the application layer of the in-home PLC communication protocol.

3.2 General characteristics of the expert system

It is important to give a brief description of the application in which the fuzzy inference engine is used. The inference engine operates in association with a knowledge base, which contains information on the correct behavior of the appliances and on the effects of faults. The set of rules is also codified in the knowledge base.

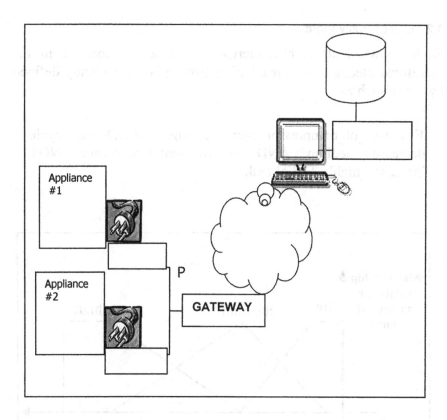

Fig. 3. System architecture

The complete workflow of the WESArd is as follows:

1. Receives compressed data from WESA devices.
2. Preprocesses data to calculate rule inputs: this task requires loads recognition, comparison of current data with historic profiles, and computation of deviations.
3. Fuzzy rules computation.
4. Diagnostic output.
5. Update of appliance profiles.

3.2 Use of fuzzy logic

Each known fault is characterized by a set of changes in the monitored electrical features. This situation is conveniently defined by a rule such as:

IF power_of_compressor increases_highly AND duty_cycle decreases_medium AND environment_temperature NOT increases_highly THEN fault1

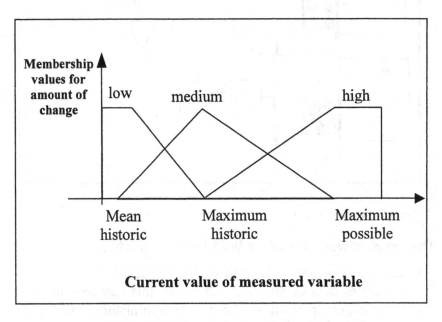

Fig.4. Definition of membership functions

The antecedents depend on the differences between current values, and regular historic values, the consequent is the diagnostic output. The use of fuzzy logic transforms the safe operating area into a fuzzy region; this approach is more correct since in most cases there is not a distinct limit between regular and faulty behavior. Moreover it allows to add quantitative information to the diagnostic output, i.e.

to what extent the fault is present. This information is fundamental for faults that imply a critic working condition and not a breakdown. The number 'maximum historic value' in chart 4, is a value resulting from learning, while the shape of membership functions is tuned in order to maximize performance of the diagnostic system on the basis of lab data, which reproduce faulty behavior in the appliances.

The semantics associated to this method for calculating antecedents is that of the severity with which a symptom has manifested itself; the severity of diagnosis is therefore a result of the severity of symptoms. In order for the consequent to be a function of all the antecedents we use the fuzzy operators "product probability", and "sum probability".

4. Benefits deriving from the use of fuzzy logic

The implementation of the fuzzy inference engine has been straightforward since it operates directly on the data structures created with queries to the database.

The use of fuzzy logic allows to obtain a quantitative information on the severity of the fault; this characteristic is fundamental since the WESArd expert system must detect faults which do not prevent the appliance from working, but which deteriorate its performance. For this kind of diagnosis an indication yes/no would not be acceptable. Information obtained through extensive laboratory tests can be conveniently expressed in terms of the fuzzy models for faulty behavior codified in the rule system.

114

References

[1] J.F. Baldwin, "Fuzzy logic and fuzzy reasoning," in Fuzzy Reasoning and Its Applications, E.H. Mamdani and B.R. Gaines (eds.), London: Academic Press, 1981.

[2] W. Bandler and L.J. Kohout, "Semantics of implication operators and fuzzy relational products," in Fuzzy Reasoning and Its Applications, E.H. Mamdani and B.R. Gaines (eds.), London: Academic Press, 1981.

[3] A. D. Gordon, "Classification", 2nd edition, Monographs on Statistics and Applied Probability, Chapman & Hall/CRC, USA, 1990.

[4] P.Arabie, L.J. Hubert, G. De Soete " Clustreing and Classification", World Scientific Publishing Co. Pte. Ltd., Singapore, 1999.

[4] Keinosuke Fukunaga, "An Introduction to Statistical Pattern Recognition", 2nd edition, Morgan Kaufmann, San Francisco, 1999.

[5] John Moubray, "Reliability Centred Maintenance", 2nd edition, Butterworth Heinmann, Great Britain, 1999.

Image Segmentation Using a Genetic Algorithm

Vitoantonio Bevilacqua[1], Giuseppe Mastronardi[1]

[1] D.E.E., Politecnico di Bari, Via E. Orabona, 4
70125 Bari – Italy
bevilacqua@poliba.it,mastrona@poliba.it

Abstract. In this paper we present a genetic algorithm-based optimisation technique for an automatic selecting of the thresholds in image segmentation, considering in a combined way, the parameters of the segmentation and the parameters of the pre-processing and post-processing operators. The design of a database based on the most important image's features to generate an accurate initial population for all the needed parameters is described. The obtained performances of the designed system are then shown qualitatively.

1 Introduction

Segmentation is an important task in computer vision, but there is no theory of image segmentation and often, the techniques are basically ad hoc[1] and emphasise one or more than one of the desired properties of the final images. The accuracy of the image segmentation process strictly depends on how different grey tone intensities have the objects to be recognised in the scene and on the presence of noise. So it is necessary to choose the segmentation algorithm and the pre-processing and post-processing operators, to determine which parameters are to be changed, how, when and how may times to do it and this task lead to a number of different choices.

In the last year several techniques have been studied for selecting the thresholds in the grey tone intensity histogram [2][3][4].

In this work we have just used two classes of smoothing [4] pre-processing filters, i.e. the third order low-filter (FPB) with b parameter in the integer range from 1 to 4 used for a number of times ($nFPB$) up to 3, and the median filter (FM) with cross (if

$boolX$ is 1) and rectangular (if $boolX$ is 0) neighbours and with an order ($ordM$) in the range of odd numbers from 1 to 7 and repeated nM times up to 4. The segmentation algorithm adopted in this work was proposed by Haralick e Wang [6] for multiple threshold selection. We have chosen this algorithm instead of a single threshold one in order to have more information about the images, and we have tried to use an evolutionary approach to solve the well known problem of selecting the thresholds to avoid the under and upper segmentation. Finally in post-processing we have used the operator of erosion and dilation [7], on to a 3x3 mask for a number of times $Neros$ and $Ndil$ respectively.

2 The Genetic Algorithm Segmentation

In this paragraph we present the genetic algorithm used in terms of chromosome's structure, fitness function, operators of cross-over and mutation, initial population and stop criterion. We have used a simple genetic algorithm [8], in which, step by step a new whole population is created (non overlapping populations), and then we have adopted a single-point cross-over and an uniform mutation to replace the old one.

2.1 The Chromosome

In order to design an appropriate structure of the chromosome we have investigated before which and how parameters are possibly and usefully to be changed, and then we have set for any parameters a range of variability explained in the following. Above all we can set a maximum number of threshold to segment the images by the two following ways:

1) using a low-filter on the grey level histogram with for example a 3x3 mask and evaluating the number of modes $Nmodes$ and then set $Nmodes$ like maximum value. Moreover to achieve good results from low-filter and delete the peaks not related to actual modes, we have repeated them $Nmeans$ times, with $Nmeans$ variable from 10 to 40 with step 10. Higher values

could lead to an under-segmentation and make useless the following adoption of the post-processing operators.

2) setting a minimum gap between the thresholds in the integer range from 2 and 5. Higher values could lead to an under-segmentation and make useless the following adoption of the post-processing operators in this case as well.

It is well known the role of the parameter M and the importance of a stop criterion based on the number of edge pixels in the Hb (bright edge pixel histogram) and Hd (dark edge pixel histogram) [6].

Moreover we have discovered that sometimes is more useful to have a variability of the parameter M, usually set to 3, in the integer range from 1 to 8. At the same time the stop criterion is related to the minimum number of edge pixels in the Hb and Hd at the step t. We can then fixed, even we have realised that this is the less useful parameter, a stop criterion as a minimum percentage $ptile$ of these points in the comparison image E [6], in the range from 0.5 % to 0.20 % with a 0,5% step.

The last parameter useful for the segmentation is related to the comparison matrix of the pixels expected to be border pixels and it has been built using Roberts and Sobel filters with th threshold like a gradient estimate. On the basis of the obtained experimental results we have allowed the variability of this threshold in the range of even number between 10 and 40. The segmentation task needs five parameters: $gap, ptile, th, M, Nmean$. We have adopted the boolean $boolER$ to code which operator is used before: 1 if the dilation succeeds the erosion. We have then finally the following chromosome structure: 2 bits for b, 2 bits for $nFPB$, 2 bits for $OrdM$, 1 bit for $boolX$, 2 bits for nM, 4 bits for th, 3 bits for M, 2bits for $Nmean$, 2 bits for $ptile$, 2 bits for gap, 3 bits for $Neros$, 3 bits for $Ndil$, 1 bit for $BoolER$. As all the genes of the chromosome are integers, we have used a binary chromosome even if to avoid some complications in terms of making bit computing each allele has been coded using a char. So the whole chromosome that contains 13 genes has been coded initially using an array of 29 char and then an array of 26 char assuming to use the erosion and the dilation for the same numbers of times [11].

2.2 The Fitness Function

To design the fitness function we consider now that if we start from the input image and obtain the edge image (GI Gradient Image) evaluating the Roberts [6] operator, it is necessary to choose a threshold t so that we have a number of edge pixels nGI.

Instead, starting from the segmented image we have the edge image (BI Boundary Image) by using the Roberts operator with unit threshold, we have a number of boundary pixel nBI. We can compare one to one the pixel of the boundary image and the pixel of the edge image and then look for the threshold t that minimises the difference between nGI and nBI. To do this we could use the following formula [10][11]:

$$F_1 = \frac{GI \ and \ BI}{\min(nGI, nBI)} \ . \tag{8}$$

where nGI and nBI evaluate the number of points equal to 1 in both the images, since the threshold operation for both the matrices (edge and boundary) gives as a result 1 if the pixel is an edge. But since this kind of function could fail because often the segmented images have imprecise boundary, above all, after the erosion and dilation a different fitness function F_2 has been designed. Then if the edge of the segmented images is just shifted, it is possible to estimate this occurrence using a weight inversely proportional to the minimum distance between the pixels in the GI e BI images like in the following:

$$F_2 = \frac{(GI \ and \ BI) + \sum_{i=0}^{d-1} \frac{hist(i) \cdot (d-i)}{d} - \frac{E_1 + E_2}{2}}{nBI} \ . \tag{2}$$

where d is the size of the neighbour where to look for, $hist(i)$ is the number of pixels of the BI images at a distance i from the closest pixel in the GI image, E_1 the number of edge pixel in GI at distance bigger than d, from any other pixel in BI, $E2$ the number of boundary pixel in BI at distance bigger than d, from any other pixel in GI, $(d-i)/d$ the weight. Finally we have used the following fitness function in which we have normalised the coefficients in order to have a unitary maximum value of the fitness.

$$F = \frac{1}{2}F_1 + \frac{1}{2}F_2$$

(3)

A complete search of the threshold t value amounts to evaluate the Roberts operator with threshold variable in the range from 1 to 254. For any pixel it is necessary to calculate 4 summation, 2 modulus, and a maximum search between two numbers, and then the evaluation of nGI, (GI and BI); for this reason we have restricted the search space on the basis of the following considerations [9]. Grey level transitions from 255 to 0 in a real world images are very rare, so gradient values much close to 255, should have close to zero values in the GI grey level histogram. Then it is not useful to increase too much the threshold because the results would be a such small number of edge pixels compared to the total number, to lead to a bad fitness function value. At the same time too small threshold values, above all in presence of noise, would lead to a very large number of edge pixel, and it is useless as well. Each chromosome of each population needs of the evaluation of the fitness and then the search of the optimal threshold. Since we use the same GI image for all the chromosomes, because it does not depend on the segmentation parameters of each chromosome, it is possible to evaluate the number of edge pixels for any threshold and then reused these pre-calculated values without evaluating the same operations any time the t threshold is to be evaluated; in such a case we can use for the GI the value of t that minimises the difference between $nGI(t)$ and nBI. The gain in terms of computational time increases as the product between the number of populations and the number of generations increases. On the basis of these hypothesis we have pre-calculated $nGI(t)$ in $t = 8 + 8*k$ with $k=0,1,...11$. By doing so we have obtained good results in terms of time computing paying just a bit of memory, during the search of the optimal threshold we look for the best $GI(t)$ among the pre-calculated and not more than 7 iterations will take the algorithm to reach the final solution. The initial population (SP Seed Population) could be initialised randomly, but we have improved our algorithm with an accurate initialisation to save some iterations for any chromosome evaluation. On a 800 MHz Pentium III with 64 MB RAM, the segmentation of a 256x256 (with the 1,1,5,1,1,20,3,20,10,3,2,1 parameters) takes us 0,5 second, a population of 30 individuals takes us 15 seconds, 15 new generations take us 225 seconds. To build an initial population the idea is that similar images need similar segmentation parameters and then we built a database of well-segmented images storing the most important features of the images and their best segmentation parameters. To make an accurate comparisons among different images we have used at the same time more than one normalised features like mean, standard deviation, skewness, kurtosis, first order entropy, histogram size, number of histogram modes, size of the maximum, size of minimum valley, distance between the two largest valleys, ratio between the inertial axis, angle between the inertial axis and the x-axis, the seven Hu's momenta [10][11][12][13].

The initialisation module of the genetic algorithm evaluates the features of the images to be segmented, then compares them to the all the features of the images in the data-base to order them from the most similar to the less. Then it is possible to initialise all the genes of chromosomes at the same values of the genes just using the most similar image differently from each other choosing the first most similar ones.

To build the data-base it is necessary to load the 256 grey level bitmap of the images, set the parameters of the genetic algorithms, start the segmentation algorithm and then on the basis of the results obtained store or not store the parameters. This decision has been kept by the user evaluating the results of the genetic algorithm segmentation on the basis of the available grey level segmented images (each tone being the arithmetic mean of the two thresholds obtained) and of the edges images obtained using the Roberts filter with a unit threshold. The stop criterion used is not based only on a threshold related to the maximum value of the fitness, but involves the maximum number of evaluations as well, because owing of the difficulties to obtain a good match between the GI and BI in all the test we have run successfully we have obtained good results with fitness value about 0,70 or 0,75 and sometimes small value of fitness does not produce bad results.

3 Experimental Results

Since the objective of this work is to implement an automatic system to segment different class of images we will consider a series of images, each representative of a class, like input of the system, and then we will evaluate the evidence of their segmented copies varying any characteristics parameters of the adopted genetic algorithm.

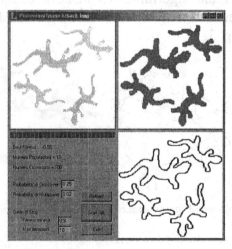

Fig. 3. In the four panels are shown in a clockwise the input image, the segmented image, the edge image of the segmented image, in order to have an alternative picture to evaluate the results of the segmentation, and the parameters of the genetic algorithm (the probability of crossover, the probability mutation, both two the stop criterions the minimum value of the fitness and the maximum number of generations, the dimension of the population, the number of actual generations and the value of fitness of the best fitted chromosome belonging to the last generation).

We begin with the 'lizback.bmp' image, 199x202 pixels, shown in the fig 1,2,3. It represents a grey tone images with low contrast and

122

with a four lines histogram. Each animal presents gulfs and irregular peninsulas, and a not monochromatic colour as well, so that it is not a trivial task to detect the edges. Despite of these difficulties the system returns a two colours image (having chosen a single threshold) in which the shapes are correctly detected and the empty zones inside the legs have been completed using the pre-processing filters or the morphological operators of erosion or dilation, but, in any case, in an automatic way. In this case, even if with a low number of generations (5-6) a good result is reached. Furthermore the probabilities of crossover and mutation have a little influence on the result. In the fig 2. is shown the best solution of first generation. Since the initial population has been created starting from the t-uple in the database (the genetic operators have not been still used), the obtained segmentation is a measure of the help that the database could give. Accordingly, the bigger is the number of the t-uple in the database similar to the image being segmented, the faster is the convergence of the algorithm. To make a comparison the same image but segmented with a traditional single threshold algorithm is shown in the fig. 3.

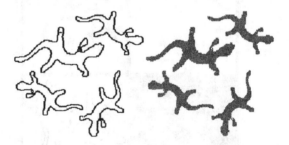

Fig. 4. The best solution of first generation of the genetic algorithm

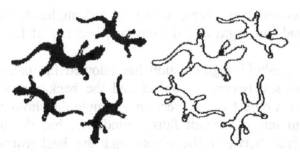

Fig. 5. The 'lizback.bmp' image but segmented with a traditional single threshold algorithm

We consider now three 64x64 images, acquired with not uniform illumination and corrupted by noise. All the images present a histogram with a peak strongly predominant in correspondence of the object and a number of secondary peaks in correspondence of the background. The results shown in the fig 4 are obtained with the 50 chromosomes, 0.25 probability of cross-over, 0.02 probability of mutation, 10 maximum Number of generations, 0.9 least Fitness. The obtained fitness values are respectively $f_1=0.748$, $f_2= 0,887$ and $f_3= 0,786$. For the second and the third image of the fig.4 the acquired result seems to be the best obtainable, since the genetic algorithm does not longer evolve increasing the population of individuals and the number of maximum iterations. Furthermore the genetic algorithm has been run a limited number of times (12 for each image and for each configuration of parameters). All the three

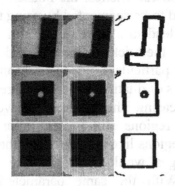

Fig. 6. The input image, the segmented image and the edge image obtained.

124

images present bad edges, in the up left angles for the first two images and low right and for the last, because of the not uniform lighting.

A multi-threshold segmentation has allowed the understanding of the shadows between the object and the background, allowing a more correct object edge detection. A single threshold segmentation obtains an image without false boundaries, but the bigger is the shadow area between the object and the background the more imprecise the contours.

Fig. 7. The results obtained increasing the number of iterations up to 20.

Moreover in the first image increasing the number of iterations up to 20, better performances are achieved, no false boundaries are present in the fig.5 and shadow zones are still detested. In this case in fact we have obtained $f_1= 0.915$, and since $f_1>$ [fmin]= 0.9, one of the two stopping criterions occurs and the genetic algorithm does not reach the maximum number of iterations taking 16 generations, 6 more than in the previous run.

Finally we consider now the image 256x256 'House.bmp' that presents a multimodal grey level histogram. This image has got areas with different characteristics: the house is made using bricks and then it is not good to be segmented by using thresholds; a simple segmentation would lead to a large number of very small regions, likely coincident with the bricks or the tiles and not the whole body of the house with its particulars. For this reason the pre-processing filter and the operators of dilation and erosion are of the fundamental importance to decreasing the contrast between the bricks to eliminate the small regions still present after the segmentation. Finally the brighter regions like the gutter and the windowsill having sharp lighting changes would lead to false boundaries in the segmented image. With the same parameters of the previous example we have obtained a good result in fact the body of the house is not fragmented for the bricks texture, the roof at the contrary is partially fragmented, because the texture of the tiles is bigger than the bricks, the sky is properly segmented, the window lost its superior line owing of the shadow, the shadow of the roof on the wall has been segmented like a different and homogeneous region.

Fig. 8. Obtained results on a multimodal histogram image.

4 Conclusions

In this paper we have cast the problem of selecting the thresholds in image segmentation as a cost minimisation problem. The use of a database of pre-segmented different classes of images, to initialise the new population in the evolutionary strategy is shown to perform very well, and the results of the segmented images obtained are qualitatively good.

References

1. V.Bevilacqua, A.Sappa, M.Devy: "Improving a genetic algorithm segmentation by means of a fast edge detection technique", ICIP 2001, Greece (2001)
2. J.S.Weszka, A.Rosenfeld Histogram Modification for Threshold Selection IEEE Transaction on SMC,. 38-52, (1979)
3. S.Wang, R.M.Haralick Automatic Multithreshold Selection Computer Vision, Graphics and Image Processing,.46-67, (1984)
4. J.S.Weszka A Survey of Threshold Selection Techniques Computer Graphics and Image Processing,. 259-265, (1978)
5. P. Zamperoni, Metodi di elaborazione digitale delle immagini, 1990, Masson

6. R.M.Haralick, L.G.Shapiro Image Segmentation Techniques Graphics and Image Processing,. 100-132, (1985)
7. R.M.Haralick, S.R. Sternberg, X.Zhuang, Image Analysis Using Mathematical Morphology,. Transactions on PAMI,432-549, July,(1987)
8. Michalewicz, Z.: Genetic Algorithms + Data Structures = Evolution Programs. 3rd edn. Springer-Verlag, Berlin Heidelberg New York (1996)
9. V. Bevilacqua, Modelli Evoluzionistici per la Visione Automatica, Ph.D. thesis (2000)
10. B.Bhanu, S.Lee, J.Ming Adaptive Image Segmentation using Genetic Algorithm IEEE   Transaction su SMC,. 1543-1567, 1995
11. P.Zingaretti, A.Carbonara, P.Puliti Evolutionary Image Segmentation Proceedings ICIAP Firenze , (1997)

How Fuzzy Logic Can Help Detecting Buried Land Mines

E. Gandolfi, P. Ricci, M. Spurio, S. Zampolli

Department of Physics University of Bologna,
INFN sezione di Bologna

Keywords: Fuzzy Logic, buried land mines, nuclear physics techniques, thermal neutron analysis

Abstract. One of the most relevant problems of the recent wars is given by the enormous amount of buried land mines disseminated over the country when wars are over. Lots of methods for cleaning a minefield are developed all over the world. Most of these methods use a specific sensor and a decisional system to recognize the mine presence. This work describes how Fuzzy Logic can help to build a software tool that performs the decisional step able to reveal the mine presence by analysing the output signal from a specific mine sensor built by the EXPLODET project of the Italian INFN. The probability to reveal a mine has to be greater than 99.6%. This constraint is required by the United Nations to tailor the objectives of all R&D initiatives in this field [1]. The results obtained proved useful in building a simple and efficient method to face this problem. The two methods presented in this paper are meant to be used together with other detection ones, such as metal detection, to confirm the result of the other instruments in the conditions in which these ones don't reach the requested precision. It is well known that only a combination of different sensor systems can fulfill the requirements of humanitarian demining activities.

1 The buried land mine problem

The International Red Cross [2] reports the presence of more than 120 millions of buried land mines worldwide and they represent a threat to the population for a long time after the end of the military conflicts.
The detection and removal of buried land mines is a complex and very dangerous task. The existing methods used to clear a minefield

are very slow, expensive and not completely reliable, causing death and injury of many specialists involved in humanitarian activities.

Considering these limits, the need for a new detection method based on a completely different approach is evident. The aim of the INFN EXPLODET project is the realization of a land mine detector using advanced nuclear physics techniques. This method can efficiently reveal specific chemical elements, in relatively short time [3]. Considering the various aspects of land mine detection, an efficient and available technique at present is the detection of γ-rays from the nitrogen de-excitation. The nitrogen, contained in every explosive in concentrations greater than 10%, can be excited using thermalized neutrons.

This method uses a simple and easy way to implement data analysis, based on the Fuzzy Logic, completely automated. The goal is to obtain a detector with high reliability, combined with short detection times and low cost.

2 The Thermal Neutron Analysis (TNA) method and apparatus

The TNA is based on the identification of the γ-rays emitted by buried land mine nitrogen, after the capture of thermal neutrons. The neutrons come from an external source. After the neutron capture, the nitrogen nucleus in the explosive shifts into excited and short living states. The following decay back to the ground state happens with the emission of γ-rays with characteristic energy of 10.8MeV (full energy peak) and 10.3MeV (first escape peak). Since there are no other peaks in this energy interval, the presence of an abnormal activity over the background is an indicator of high concentration of the nitrogen, used in the production of the explosive material.

The TNA-devices implemented in EXPLODET, are based on a thermal neutron source, a gamma ray detector, and a Multi Channel Amplifier (MCA) interfaced to a PC running decisional software (Fig. 1).

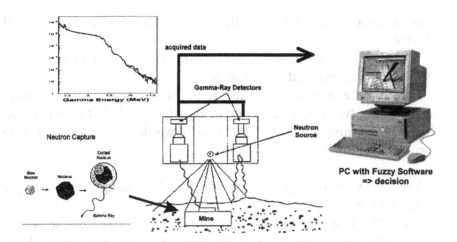

Fig. 1. Schematic layout of the TNA technique used by EXPLODET. The Fuzzy Logic is used to build a critical step of the logic process able to take the decision about the presence of a land mine.

The neutron source consists of ^{252}Cf, which decays by alpha emission (96.91%) and spontaneous fission (3.09%). In each fission event, 3.7 neutrons are emitted on average. These neutrons have energies between 1 and 12MeV, with a peak at E_n=2MeV. Because the energy required for nitrogen capture is about 0.025eV, a moderator is required to thermalize the neutrons. [4] At present the neutron source has an intensity of about 2×10^5 fission/s.

Our solution for gamma ray detection is a CsI(Tl) scintillator coupled to photodiodes. The γ-ray spectrum is collected through a MCA. This detector is highly efficient in the 10MeV energy range and it has a reasonably low cost.

3 TNA spectra analysis methods

Several problems arise when dealing with γ-ray spectra produced by TNA. The major issue is the shift of the energy scale due to intrinsic properties of the acquisition system. In fact, it is impossible to know the energy corresponding to a single channel of a MCA with the required precision. In on-field applications, various effects can produce a shift of more than ± 20 channels on 512 channel devices.

We outline that the number of channels involved to reveal the 10.8MeV peak is about 30 channels.

Furthermore a real time calibration of the apparatus, generally made using two calibrated sources, is not possible because it is too difficult to implement on field, and it would require a very long exposition time.

We studied two ways to solve this problem using a Fuzzy Logic software tool.

- The first method is based on some considerations to reveal the two nitrogen peaks directly analysing the spectra shape produced by the MCA.
- The second method faces the problem of the MCA energy scale calibration using Fuzzy Logic for this specific purpose. Then the mine presence is directly connected to the statistical counts in the region related to the two nitrogen peaks.

4 First approach: Fuzzy system for the revelation of the 10.8MeV and 10.3MeV peaks

The Fig. 2 shows the two γ-ray spectra obtained by the MCA when the apparatus explores a clean field (dotted line) and a zone with a buried mine (continuous line). The zoom emphasizes the 10.3 and 10.8MeV zone where the energy peaks related to the explosive counts are somewhat greater than the background counts. The basic idea of the EXPLODET project is to analyse these data and to confirm the presence of a mine in a short time when this difference is greater then a prefixed threshold.

We can also notice that the shift of the energy scale of 20 channels can modify the spectrum and generate a possible mistake if one analyses only this region. To improve the precision of the decisional step and to reduce the decision time we have chosen to take into account, using the Fuzzy Logic, how the 10.3-10.8Mev region and other zones evolve in time.

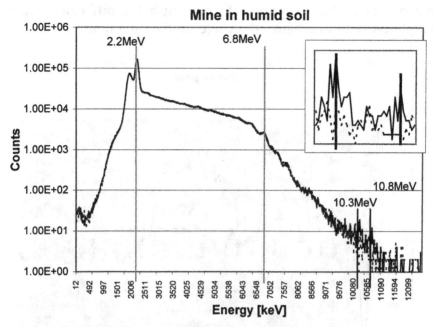

Fig. 2. Spectrum obtained by the EXPLODET detector in 500s acquisition time. Dotted line is related to the background. Continuous line is related to 800g buried explosive. You can see the first 2 peaks used for the energy calibrations of the detector and the 2 pecks related to the nitrogen, these peaks are evidenced in the upper right box.

In this approach to the problem the first step involves the collection of the background spectrum in a zone where we suppose there isn't a mine. Then we explore the zone where other instruments have detected a possible mine to confirm or not its presence.

In Fig. 3 we plot the difference between the explosive counts and the background counts in a region nearly more extended than the interested one and for a time longer than the one shown in Fig. 2. We chose 4 arbitrary groups of channels with the same features for all the different spectra collected by the EXPLODET team: 2 external groups where the count never increases and 2 internal groups where the count increases when there is a mine.

The Fuzzy system should give a high affordable answer in a time as short as possible. We define a time quantum of 1000s and every time quantum the system checks the new acquired spectrum and gives an appropriate answer. The formula (1) defines the $Nj_{delta}(Tn)$

132

parameters as the indicators of how much the difference counts (explosive minus background) evolves in time at n-time.

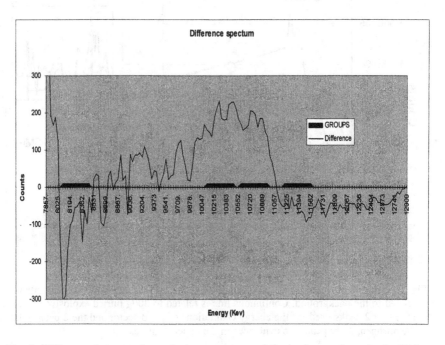

Fig. 3. Difference between the explosive spectrum and the background one in the 10.3-10.8 MeV energy zone after 10000s. The four groups used for the Fuzzy system input variables are highlighted.

$$Nj_{delta}(T_N) = \frac{\Delta Cj(T_N) - \Delta Cj(T_1)}{\sigma_{\Delta Cj(T_N)} + \sigma_{\Delta Cj(T_1)}} \tag{1}$$

where j is relative to the 4 groups and the $\sigma_{\Delta Cj(Tn)}$ is the standard Poissonian error at the T_N instant.

The $Nj_{delta}(Tn)$ parameters represent our 4 input Fuzzy variables.

The output variable of the system is represented by a real number in a 0-1 interval with the following features:

explosive absence => output $\cong 0.0$;

uncertainly => output $\cong 0.5$;

explosive presence => output $\cong 1.0$.

Instead of building the Fuzzy set parameters and rules by logical consideration extracted by experience, we decided to use a training phase using the software tool AFM (an Adaptive Fuzzy Modeller produced by SGS-Thompson to implement Fuzzy systems in WARP Fuzzy processors [5]). When we started we had only a little number of original spectra produced with and without mine presence. To increase this number we have simulated 400 spectra starting from the original data by a Montecarlo randomisation process. So we obtained 200 spectra at different time (from 1000 to 10000s) to use as training and 200 as a test for the system.

In order to use these spectra as learning examples we had to assign an output for every one used as input.

We chose the following criteria.

- The value 1 was given under two condition:
  o the data set is related to a mine presence spectrum;
  o at the time when N_{delta} for the internal groups (the groups in the 10.3 10.8MeV zone) exceeds the numerical value of 4 and the two external groups are confined in the 0-1 interval. This is equivalent to 4-times the statistical error giving a 99.6% error proof system. This occurs in the T_7-T_8 instant in most of the cases.

- The value 0 was also given under two conditions:
  o the data set is related to a explosive absence spectrum;
  o at the T_7-T_8 instant when all the 4-N_{delta} groups were in the 0-1 interval that is the counting difference in the 10.3 10.8 MeV zone has no statistical meaning.

- The value 0.5 was given in the initial instant T_2 and in all the other cases.

Different numbers of Fuzzy sets and different shapes for the 4 input variables were tested. The best results were obtained with 3 gaussian Fuzzy sets for the all the variables. With these conditions in the training phase AFM converged rapidly, and after 300 iterative learning epochs a working optimised Fuzzy System was produced.

The system was tested using 200 new input-output patterns. The results, with the spectra in presence of explosive, have shown that energy MCA shifted and stretched with a maximum of 5 channels didn't create any problem to the output and the correct answer in a short time (instant T_6-T_7) was obtained choosing a suitable threshold for recognizing the explosive presence. But with 10 channels distortion the system failed to reach an appreciable output value near the threshold and the uncertainly value remained.

In case of explosive absence otherwise the correct answer of 0 was always given.

Since the EXPLODET apparatus could have a ±20 channels error on on-field application this means that the calibration problem cannot be solved in such a way and a new approach is necessary to overcome the problem.

5 Second approach: Fuzzy system for the calibration of the energy scale

5.1 Analysis of the problem

In this new approach we have built a Fuzzy system for solving the MCA calibration problem. For calibrating the spectra we need two reference peaks well recognizable and at well-known energy values.

To improve the precision of this calibration, we have chosen the two reference peaks as far as possible from each other. The first one is the 2.2MeV peak produced by thermal neutron capture from hydrogen. The second is a 6.8MeV peak, which is not caused by any specific element but rather by the convolution of various activities. Both peaks are present in every spectrum acquired with the EXPLODET detector (Fig. 2). The MCA energy is assumed to be a linear function of the channel number, allowing the extraction of the parameters of this function when the exact channel numbers of two reference energies are known.

The 2.2MeV peak is easy to identify (it is always the global maximum of the gamma ray spectrum). The 6.8MeV peak is much more difficult to recognize. Several classic analytical approaches were considered, but the presence of many peaks in the 7MeV region and the difference between various acquired spectra (mine in dry and humid soil, mine in PVC-pipeline, mine in metallic shells, etc.) requires an adaptive and flexible method.

Once the energy scale has been calibrated and the position of the peaks is exactly known, the recognition of any abnormal activity in the 10-11MeV range is possible with a classical statistical counting. Furthermore, analyzing the spectrum we observe that it can be fitted with a second order polynomial. Then this property will be used to determine the no-mine reference spectrum, instead of collecting a background spectrum as first step of the mine detection as described in the previous method.

5.2 The Fuzzy system basic idea to detect the 6.8MeV peak

From Fig. 2 the expert can notice (from left to right) after the 2.2MeV peak a quite linear descending slope for about 200 channels (channels 110-320). Few small peaks are then present in the following channels. This is true not only for the two spectra in Fig. 2 but also in all spectra collected by the EXPLODET group. These peaks have somewhat different features from the 6.8MeV peak we are looking for.   The 6.8 MeV peak is:

- almost in all cases, higher with respect to the neighbouring peaks;
- in a region of the spectrum with a higher average slope.

These two features, used by the expert, have helped us to define the input variables of the Fuzzy system able to automatically detect the 6.8MeV peak with the needed precision in a very short time.

Some free parameters, as the intensity of the neutron source and the exposition time, modify the number of the events for each energy interval of the spectrum. In order to have similar conditions, the collected data are normalized to the maximal count, corresponding to the number of events in the channel of the hydrogen peak (2.2MeV).

Due to the exponential decay of the counting rate along the energy scale, the logarithm of the normalized spectrum is calculated to obtain linearization, and from now on every considerations relate to this quantity.

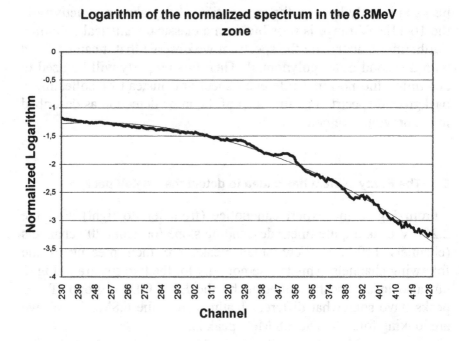

Fig. 4. The figure shows the energy spectrum region (channels 230-422 of the MCA) used to identify the position of the 6.8 MeV channel. The thin line is the result of a 2nd order polynomial fit.

5.3 Features of the Fuzzy System

The first input variable (D) of the Fuzzy System is the difference between the MCA counting related to the spectrum and the quadratic fit of the spectrum in a selected region (channels 115-315). This choice emphasizes the peaks of the spectrum, as shown in Fig. 4. In the case shown in Fig. 5a, the 6.8MeV peak is the global maximum in the considered range, but this is not true for all the measured spectra. In fact we collected spectra (in particular from explosive material inside metallic and PVC pipelines), where other peaks in the region are more evident.

The second input variable (Z) represents the slope of the spectrum. To define this variable we observe that the tangent line of the experimental spectrum changes significantly point by point. The average value on a large enough number of channels is more regular and approximates a negative slope straight line. We define the function Z in a channel as the average value of the tangent of the spectrum over 33 channel values before and after the considered channel.

The number of channels was chosen after many trials and over different types of spectra. Phenomenologically, we found that for a large number of experimental spectra the 6.8MeV line corresponds to the channel of the MCA where the value of the function Z is near the −0.015 value.

For each channel with input variables D and Z, we define an arbitrary output value proportional to the probability for the considered channel to contain the peak. The point with maximum output value corresponds to the position of the 6.8MeV channel.

To determine the Fuzzy set membership functions and rules we used the AFM software described in the previous paragraphs. In this case we had the possibility to use a larger type of original data spectra corresponding to the different operating conditions (dry soil, humid soil, PVC pipeline and others). Starting from these experimental conditions we have built 1200 training set using a Montecarlo randomisation process. The best results were obtained with 3 gaussian Fuzzy sets for the first input variable and 5 gaussian Fuzzy sets for the second one (15 Fuzzy rules in total).

Testing the system with 1200 new randomised spectra we have found that it is possible to calibrate the MCA with a precision of ±1 channel on spectra shifted and stretched up to ± 50 channels, much more than the 20 channels required by the EXPLODET specifications.

5.4 The decisional method to detect buried land mines

The global decisional software, written in Borland C++ on Windows platforms, runs through the following steps that are sequentially processed each spectrum update:

1. processing the spectrum data and evaluating the input variables of the Fuzzy system;
2. calibrating the spectrum using the Fuzzy system based on the parameters and rules computed by AFM;
3. performing the extraction of the channels related to the 10.8 MeV Nitrogen peak and computing the difference between the counting and the expected background without the mine.

This update is performed each time-quantum by the apparatus during exposition.

From now on we call this difference the output value R. The program also computes the Poissonian error on R. An associated 4σ, where σ is the error, matches the 99.6% probability required to notify the mine presence.

Using this software tool we have analysed spectra collected in different soil environment with and without explosive. The output value R normalized to the expected average background after an exposition time of 7500s is shown in Fig. 6.

For a given soil conditions in absence of buried land mine, we have an average positive or negative value of R. The fact that both R is not zero and the errors are different from point to point depends on the particular composition of the soil, which modifies the background expectation. In spite of these differences the value of R in presence of a mine has a great positive value respect to the background value in all the measured spectra.

In the first four cases of Fig. 6, the presence of the explosive is revealed if the normalized counting rate is at least 4σ greater that zero. In the other two cases (characterized by the presence of metals around the explosive) it is necessary that the normalized counting rate is 4σ greater to the threshold level of 0.3. In all cases shown in the figure, the presence of the explosive was revealed by our technique.

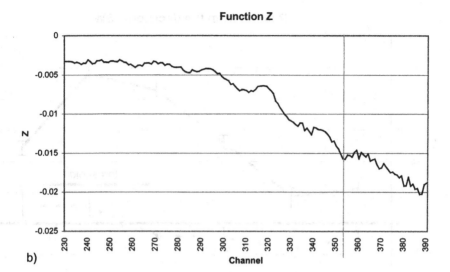

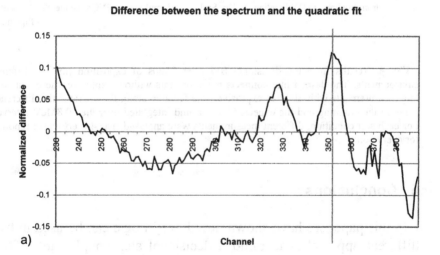

Fig. 5. a) the variable D (the difference between the normalized spectrum counts and the result of the 2nd order polynomial fit). **b)** the variable Z (the derivative average of the spectrum) is represented vs. the channel number.

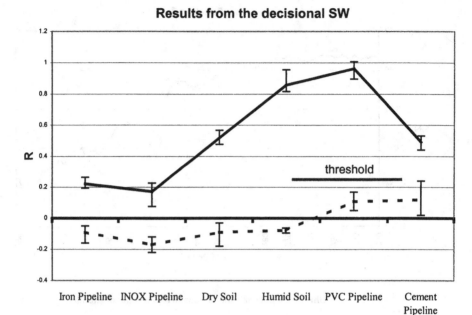

Fig. 6. Results from the decisional SW after 7500s of exposition time in different environments. The dashed line connects measurements without explosive, the continuous line measurements with 800g of explosive. The R value along the y-axis is the difference between the counting and the expected background integrated over the 10.8MeV interval. For each soil condition, different measurements were analysed. The 4 σ errors are reported for every point.

6 Conclusions

In this paper we have shown how Fuzzy Logic can be used in two different approaches to build a decisional step coupled to a TNA mine detector. The first method presented extracts the features directly from the input data set and gives an answer with regard to the various detection errors of the system. This method returns high precision result only if the MCA (multi channel amplifier) energy calibration error is limited. The second method overrides this limit. In this case the Fuzzy Logic is used to detect the 6.8MeV peak and therefore to automatically calibrate the energy scale of the MCA. We also outline that this method to calibrate a MCA based on Fuzzy

Logic could be used in many other applications and then acquires a technological relevance. We notice that both these methods take quite a long time. This time will be substantially reduced using a neutron source 10 times as powerful as the present one, thus reducing the decision time of the same factor.

References

1. "International Workshop on Localization and Identification of APM", Report EUR 16329 EN, 1995

2. International Red Cross sources at Website http://www.icrc.org/eng/mines

3. J.E. McFee and A. Carruthers, "Proceeding of the Conference on Detection and Remediation Technologies for Mines and Mine-like Targets", Orlando 1996, SPIE Vol. 2765

4. G. Viesti et al "The EXPLODET project: advanced nuclear techniques for humanitarian demining" Nuclear Instruments and Methods in Physics Research A 422 (1999) 918-921

5. AFM manual at ST Website: http://eu.st.com/stonline/books/pdf/docs/6087.pdf

6. Sugeno, M., "Industrial applications of Fuzzy control", Elsevier Science Pub. Co., 1985

Neuro-fuzzy Filtering Techniques for the Analysis of Complex Acoustic Scenarios

Rinaldo Poluzzzi, Alberto Savi

ST MICROELECTRONICS, Via C.Olivetti 6, Agrate Brianza, Milano, ITALY
rinaldo.poluzzi@st.com, alberto.savi@st.com

Abstract. A neuro-fuzzy based filter for acoustic filtering is described. In order to perform temporal filtering, a proper architecture was developed exploiting a buffer memory. Subband analysis was used. This led to an architecture composed of a QMF filter bank, a neuro-fuzzy network for each subband and a reconstruction QMF filter bank. Many simulation results are described relative to signals captured from real acoustic scenarios.

1 Introduction

Pure tones are rarely found in nature. More generally, we find sounds of arbitrary power spectrum density that are mixed one with the other. As an example, we can take a signal composed of voice mixed with music or other noises coming from a town environment: noise coming from car engines, noise coming from the crowd, etc. Our brain detects one sound from the other in a short time. In the field of signal processing various techniques are widely known for enhancing signals affected by white noise. The noise can be colored: flicker noise, with $1/f$ power spectral density is the most widely known type of non white noise. In the most general situation we have signals of arbitrary shaped spectral power density mixed one with each other. This situation is in the domain of "auditory scene analysis" and the mixed signals are called "acoustic scenarios" [1].

To solve the problem of separating a component from the other in a complex acoustic scenario, a neuro-fuzzy network was used. An architecture was developed comprising a buffer memory and a unit for signal reconstruction. The neuro-fuzzy network, with the buffer memory for temporal processing, is a neuro-fuzzy FIR. These units have the aim to perform temporal processing. Median filtering or order statistic filters are widely used for noise reduction. The median filter can smooth out impulsive noises, but cannot smooth Gaussian noises. The order statistics filter can smooth the gaussian noises, but cannot preserve sharp edges of the signal and vice-versa. As proposed in [2] a neuro fuzzy FIR, with taps calculated by means of the neuro-fuzzy network, is a solution for extimating non stationary signals with abrupt changes (acoustic signals, images).

The network performed well on signals affected by white noise. In order to enhance the performance of the network, an architecture of networks based on subband analysis was also simulated. The signal was decomposed by means of a QMF filter bank, the signal in each subband was filtered by means of the neuro-fuzzy network, and then all the signals coming from the various subbands were recomposed by means of the complementary reconstruction QMF filter bank. The rationale of such an architecture is that if a colored noise is to be filtered, within the single subband the noise is whitened, and this leads to better performance for the network.

2 Architecture of the neuro-fuzzy FIR

The neuro-fuzzy based filter solves a fixed learning problem. This means that the network generates a model of the acoustic scenario based on a couple of signals: a source signal and a target signal. During training the adjustable weights of the network are updated in order to optimize the distance between the output signal and the target signal, on the basis of a well chosen metric (or fitness function). When the distance between the output signal and the target becomes smaller than a prefixed threshold or when a fixed number of generations of the optimization algorithm have occurred, the training is stopped and the network is memorized. If the network has good generalization properties, it is able to enhance signals of the same spectral composition as the training signals.

Fig.1 describes the general architecture of the neuro-fuzzy network based filter. The filter operates on a $2N + 1$ sample window (1). Three features of the signal are extracted from this window. These features are the neuro-fuzzy network inputs. After processing by the neuro-fuzzy network (2), the output signal is evaluated in block (3) which generates the weighted sum of the window samples; the weights are the output of the network. The output signal is compared with the target and the distance between the target and the output signal is evaluated on the basis of the fitness function. After this, there is a new modification of the network weights and a new fitness evaluation

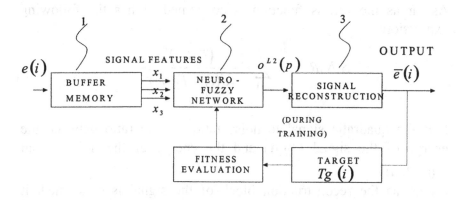

Fig. 1. General architecture of the neuro-fuzzy network based fiter

The expression of the three features of the signal is reported below:

$$x_1 = dist\ (i) = \frac{|i - N|}{N} \tag{1}$$

$$x_2 = diff\ (i) = \frac{e(i) - e(N)}{\max(diff_i)} \tag{2}$$

$$x_3 = diff \_ av(i) = \frac{|e(i) - average|}{\max(diff \_ av_i)} \tag{3}$$

with

$$(1 \leq i \leq 2N+1)$$

In these expressions $e(i)$ is the generic sample in the window.

As far as the fitness function is concerned, it has the following expression:

$$S.N.R. = \sum_{i=1}^{N \_ samples} \frac{(Tg(i))^2}{(\overline{e}(i) - Tg(i))^2} \tag{4}$$

It is the quadratic signal to noise ratio i.e. the ratio between the energy of the signal $(Tg(i))$ and the energy of the noise signal $(\overline{e}(i) - Tg(i))$.

As far as the reconstruction block of the signal is concerned, it carries out a weighted sum of the samples in a window; the weights are the output of the third layer neuron of the network:

$$\overline{e} = \frac{\sum_{p=1}^{2N+1} o^{L2}(p) \cdot e(p)}{\sum_{p=1}^{2N+1} e(p)} \tag{5}$$

3 Neuro-fuzzy network

The neuro fuzzy network is a three layer neuro-fuzzy perceptron. As far as the neuro-fuzzy topology is concerned, see fig. 2 The first layer performs fuzzification of the input variables: in this layer the degree of membership of the input features to a generic fuzzy set is evaluated. The weights of the first layer are respectively the mean and the variance of the Gaussian curves that represent the fuzzy set membership functions. These weights are updated during the network training. The outputs of the first layer are calculated according to the formula:

$$o_m^{L0} = \exp\left(-\left(\frac{x(l) - W_{mean}(l.m)}{W_{var}(l,m)}\right)^2\right) \tag{6}$$

(where $W_{mean}(l,m)$ and $W_{var}(l,m)$ are the mean value and the variance of the gaussian curve).

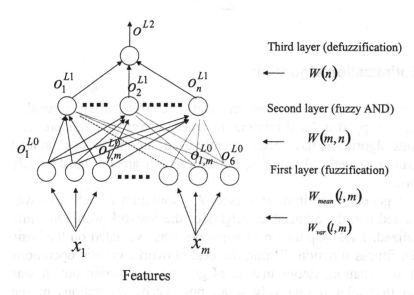

Fig. 2. Neuro-fuzzy network topology

The second layer performs fuzzy AND. The weights connecting the first layer to the second are randomly initialized and are not updated.

The outputs of the second layer are calculated according to the formula:

$$o_n^{L1} = \min_m \left(W(m,n) \cdot o_m^{L0} \right) \tag{7}$$

The third layer is the defuzzification layer and gives a discrete output value. The defuzzification method is that of the center of gravity. The third layer weights are updated during the network training. The output is calculated according to the formula:

$$o^{L2} = \frac{\sum_n W(n) \cdot o_n^{L1}}{\sum_n o_n^{L1}} \tag{8}$$

4 Optimization algorithm

As well known, it can happen that backpropagation algorithm remains trapped in local minima. In order to overcome this problem various algorithms have been recently used: among these simulated annealing, genetic algorithms, random search are the most widely known.

First a genetic algorithm was used. A population of networks was generated and the adjustable weights of the network were randomly initialized. Each population of networks was evaluated on the basis of the fitness function. Within the best networks genetic operations such as mutation, ricombination of genes were carried out. It was found that after a relatively small number of generations of the genetic algorithm the fitness function did not become significantly better.

Also an algorithm of random search was used. A population of randomly chosen solutions was generated and each solution was evaluated on the basis of the fitness function. If a candidate absolute maximum was found, an algorithm of refined search was exploited. This allowed evaluation of the behavior of the fitness function near the candidate maxima.

It was found that, for the genetic algorithm, each generation took a longer time. On the contrary, for the random search algorithm, a large number of generations was necessary, each generation requiring a shorter time and a smaller number of calculations. If a large number of variables function has to be optimized, like in this case, the random search algorithm can be faster.

5 Subband analysis

Subband analysis was used in order to enhance the performance of the system. As said in the introduction, this led to an architecture in which the input signal was decomposed via a filter bank in various subbands (two, four, eight according to various situations), the signal within each subband was filtered by the neuro-fuzzy filter, and then all the signals coming from the various subbands were recomposed by the complementary reconstruction filter bank.

As far as the design of the filter bank is concerned, see [3] and [4]. The filter bank described in the first paper generates non stationary wavelet packets with a good frequency localization. The filter bank described in the second paper generates a class of minimum bandwidth, discrete time orthonormal wavelets. The wavelets were generated via the filter bank framework and optimized by using the global optimization technique of adaptive simulated annealing. The objective function to be minimized is the averaged normalized bandwidth of the wavelets over all the scales as obtained by the filter bank structure. The discrete time wavelet transform (DTWT) of a signal $x(n)$ is defined as

$$y_k(n) = \sum h_k(m) x(I_k n - m), 0 \le k \le L \qquad (9)$$

with

$$I_k = 2^{k+1}, \quad k = 0,..., L-1, \quad I_L = 2^L.$$

The $y_k(n)$, $k = 0,..., L$, are called the wavelet coefficients of the signal , and the $\{h_k(n)\}$ can be viewed as a bank of $L+1$ filters with $h(n) = h_0(n) = (-1)^n g^*(N-n)$, where $g(n)$ is the wavelet defining filter of length N. The DTWT of $x(n)$ is obtained by passing $x(n)$ through a binary, tree-structured analysis filter bank, and $x(n)$ can be recovered from its wavelets coefficients via

$$x(n) = \sum_{k=0}^{L} \sum_{n} y_k(m) \eta_k(n) \tag{10}$$

The wavelet bases are related to the synthesis filters by

$$\eta_{km}(n) = f_k(n - 2^{k+1}m), \quad k = 0,...., L-1 \tag{11}$$

$$\eta_{Lm}(n) = f_L(n - 2^L m) \tag{12}$$

See [4] for the functionals to be minimized in order to obtain minimum bandwidth of the transform for low-pass, band-pass and high-pass sequences. Wavelets with a good frequency resolution are well suited for the problem to be solved, which is a noise reduction problem. Subband analysis whitens non-white noise within a single subband, and leads to good noise reduction also in subbands where the ratio of signal energy to noise energy is small.

6 Simulation results

The following simulations have been carried out:

- filtering of artificial signals with white noise;
- filtering of real signals with white noise.

The artificial signals were sinusoids or square waves. The real signals were voice fragments. Since the network solves a fixed learning problem, for each simulation two steps were carried out:

- the network was trained on a couple of source-target signals;
- after the training the network was tested on new signals similar to the old ones.

If the network performed well on these new signals, it had good generalization properties.

Other simulations were carried out on signals affected by structured noise and on signal mixtures. In all these simulations subband analysis showed better results with respect to simple neuro-fuzzy filtering. The simulations can be grouped in four series:

- filtering of voice fragments with pink noise (male voice);
- filtering of voice fragments with pink noise (female voice);
- filtering of voice fragments mixed with music (symphonic music);
- filtering of voice fragments mixed with music (rock music).

For acoustic signal capture a sound card Yamaya Audio was used. A 11025 Hz sampling frequency was chosen. All the acoustic files were nearly 6 sec long and composed of about 77000 samples.
In all the simulations a good S.N.R. enhancement has been obtained (up to 15 dB for fragments of voice with white noise, not higher than 11 dB for structured noise). As far as listening quality is concerned, it is acceptable. In Fig. 2 are shown two fragment of voice. The first is a fragment of voice affected by white noise, the second is the fragment of voice after filtering. The fragment of voice corresponds to a vocalic sound (vocal 'e') with 44100 Hz sampling rate.

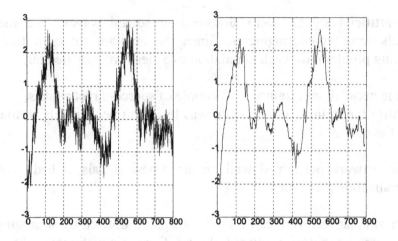

Fig. 3. Signal with noise (selection) and signal filtered (selection)

In Table 1 the results relative to filtering of files affected by white noise are reported.

Signal description	Initial S.N.R	Final S.N.R. (typical)
Fragment of voice	3.42	19-23
	13.60	50-55

Table 1. Some results relative to voice with white noise

In Table 2 some results relative to acoustic scenarios analysis are reported.

Signal description	Initial S.N.R	SNR enhancement (typical).
Male voice with flicker noise	1.58	6 dB
Male voice mixed with symphonic music	1.96	11 dB
Female voice with flicker noise	7.09	6 dB

Table 2. Some results relative to acoustic scenarios

7 Comparison with other linear methods

Various tests were carried out in order to compare the performance of the proposed architecture with other classical linear methods such as that of FIR linear filtering or that of MA. See Table 3 for some experimental results.

Signal description	Initial S.N.R	SNR enhancement (typical).
Fragment of voice (MA FILTERING)	3.76 14.20	19-20 29-30
Fragment of voice (LOW-PASS FIR FILTERING)	3.76 14.20	19-20 39-40

Table 3. Some rsults relative to linear filtering methods

8 Conclusions

A novel architecture is proposed for enhancing speech or other sounds within signal mixtures. This architecture comprises a QMF analysis filter bank for subband analysis, a neuro-fuzzy network for each subband and the complementary QMF syntesys filter bank. In all the simulation experiments subband analysis based algorithm showed better performance with respect to simple neuro-fuzzy filtering, as far as S.N.R. enhancement is concerned. Both a genetic algorithm and a random search algorithm were used for the network training. In order to achieve even better performance spatial cues due to binaural hearing could be exploited. This could lead to a multisensor system with a related neural networks architecture. This topic could the subject for further research.

References

[1] D. P. W. Ellis, Prediction driven computational auditory scene analysis, submitted to MIT in fulfillment of the requirements for the degree of Doctor of Philosophy in Electrical Engineering, June 1996.

[2] K. Arakawa, "Fuzzy Rule Based Signal Processing and its applications to Image Restoration" IEEE Journal on Selected Areas in Communications, Vol. 12, no.9, December 94.

[3] N. Hess-Nielsen, M. V. Wickerhauser, "Wavelets and time frequency analysis", Proc. IEEE, Vol. 84, No. 4, April 1996.

[4] J.M.Morris and R. Perevali, "Minimum-bandwidth discrete-time wavelets", Signal Processing, Elsevier, vol 76, 1999.

[5] P.L.Ainsleigh and C. Chui,"A B-Wavelet-Based Noise Reduction Algorithm", IEEE Transactions on Signal Processing, Vol. 4, No.5, May 96.

[6] D. Nauf, F. Klawonn, R. Kruse, Foundations of Neuro-Fuzzy Systems, Wiley, 1996.

P300 Off-line Detection: A Fuzzy-based Support System

S.Giove[1], F.Piccione[2], F.Giorgi[2] , F.Beverina[3], S.Silvoni[3]

[1] Dept. of Applied Mathematics, University of Venice
Dorsoduro, 3825/E, Venice (ITALY)
E-mail: sgiove@unive.it

[2] Dept. of Neuro-rehabilitation, S.Camillo Hospital,
Alberoni, Venice (ITALY)

[3] STMicroelectronics - AST Group
20041 Agrate Brianza (ITALY)
E-mail: fabrizio.beverina@st.com

Abstract. This article describes the implementation of a fuzzy logic based algorithm for the off-line detection of the so-called P300, an event-related potential signal (ERP) arising when a target stimulus is detected. The algorithm is based on a fuzzy-inferential engine which estimates some parameters on the averaged traces.

The resulting fuzzy inference system can analyse the averaged ERPs signals, and characterise, with some features, the most remarkable identified deflections. This characterisation can be used as a support for the physician diagnosis.

This activity may constitute the basis for the on-line P300 study, with the aim to implement a BCI device (Brain-Computer Interface), to be used as a man-machine interface and support.

1 Introduction

This paper describes a fuzzy based methodology and the related software tool for the *off-line* detection of an event related signal, the so-called P300 wave. This application provides a decision support system to the physician [1][5], for the ERPs traces detection and analysis, and it helps the characterisation of the P300 event, a relevant signal for clinical and diagnostic purposes. Those signals were induced applying the *Odd-Ball* paradigm, see par. 2, where all

156

the significant waves are sampled and averaged, while the fuzzy detection algorithm is subsequently applied (*off-line* processing).

The research activity is carried out with the collaboration of an expert neuro-physiologist, who has contributed in a significant way to the problem analysis.

The procedure initially provides for a pre-processing phase of EEG traces. In this phase some typical features are detected. In a subsequent phase, a fuzzy-inferential system [6] is applied to estimate the likeness degree of the sampled wave, compared with an ideal P300, using the features obtained in the previous phase.

Moreover, this work points out some basic items for an *on-line* P300 classification attempt [7]. The possibility of an efficient P300 recognition in the single-sweep traces (*on-line* processing), plays a basic role in a BCI development (Brain Computer Interface). Such interface would be useful as a communication device for people with severe physical disease.

2 Data and method: the *Odd-Ball* paradigm

The event-related potentials (ERPs) are recorded on the EEG scalp traces after a specific stimulation of sensorial and cognitive functionality (acoustic, visual or tactile). This technique, called *Odd-Ball*, provides for the repeated stimulation of a sensorial organ, using two different kind of stimuli, *frequent* and *rare*; at the same time, the EEG potentials of some specific derivations (Fz, Cz, Pz) are recorded by an electronic device [8].

The two types of stimuli have different occurrence percentages (*frequent* 80%, *rare* 20%), and the distribution of the *rare* events is random during the test. If the subject is trained to recognise the *rare* events, the *Odd-Ball* is called *active*, otherwise it is *passive*.

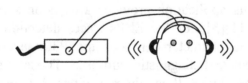

Fig. 1. Schematic representation of the *Odd-Ball* paradigm

At the end of the test, a procedure splits up the acquired traces in two distinct classes (*frequent* and *rare*), then it computes the temporal averaging for the consequent visual analysis of the neuro-physiologist.

Monitoring the averaged traces, if a peculiar deflection in the averaged trace related to *rare* events is observed, we can infer the presence of the P300 wave. Fig. 2 shows a typical P300 wave; note the amplitude peak at 356ms.

The latency and amplitude depend on the elicitation paradigm, on the subject conditions and on his/her cognitive capability[1].

The purpose of this work consists into the recognition and the characterisation of the typical deflection, using the available data from the Neuro-physiology Laboratory of the S.Camillo Hospital (Venice, Italy). In the most of the cases an acoustic stimulus is used to elicit the P300.

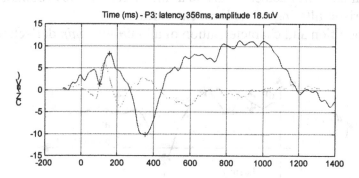

Fig. 2. Averaged ERPs traces (continuous line: *rare*; dash-dot line: *frequent*)

3 Pre-processing of the traces

The developed software analyses the traces obtained by the temporal averaging [2], implementing the recognition phase on a system based on the empirical activity performed by the expert.

---

[1] Other parameters could be sampled, but we shall consider only the most important ones, that are the latency, the amplitude and the *dRatio*, see Fig.3.

The P300 wave is defined as a wide negative and symmetrical deflection (*positive* wave), which terminates in a local minimum with a variable *latency* from 250 up to 600ms. The latency measures the temporal interval between the beginning of stimulus and the minimum P300 peak. Usually, this deflection is preceded by some *stimulus-related* components, called P2 and N2 (a local minimum and maximum), allowing to categorise the P300 wave both qualitatively and quantitatively.

In what follows, we call *deflection* any sequence composed by three local extremes (*min-max-min*), for istance, the P2-N2-P3 sequence. Thus, the pre-processing phase consists into the recognition of the deflections and their description by means of some features [3]. This activity consists of the following subsequent steps:

- low-pass digital filtering to attenuate the trace noise;
- local extremes detection inside a fixed interval (-50÷700ms);
- heuristic filtering of the local extremes;
- recognition and characterisation of all *min-max-min* deflections.

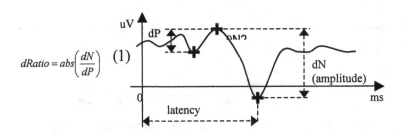

$$dRatio = abs\left(\frac{dN}{dP}\right) \quad (1)$$

Fig. 3. A *min-max-min* sequence and related descriptive features

The band of an ideal P300 is between 0Hz and 10Hz, then the digital filter allows to remove the undesired high frequency components. The characteristics of the selected filter are:

- filter type: elliptic (it minimises the digital filter order);
- pass-band: 0÷25Hz (attenuation: 0.4db);
- dark-band: >30Hz (attenuation: 26db).

The local extremes are detected using the *zero-crossing* technique applied to the prime difference of averaged traces. The heuristic filter was designed so that extremes too much close each others are discharged; a couple *max-min* (or *min-max*) is accepted if and only if both the absolute values of the difference between their amplitudes, and the differences between their latencies, are greater than two fixed thresholds.

Then, on the basis of local extremes sign, the pre-processing engine determines the *min-max-min* sequences, and extracts the features of interest [4], that are:

- *latency*;
- *amplitude*;
- *dRatio* (1) (see Fig. 3).

In the following paragraph we will briefly describe the inferential system which estimates the deflections on the basis of these three features.

4 Fuzzy sets and inferential engine

The key point of this methodology consists into the definition of the inferential rules that simulate the expert reasoning. The variables of interest (*latency, amplitude* and *dRatio*) are initially translated into *fuzzy* terms; the translation mechanism use suitable *fuzzy sets*, each of them represented by *trapezoidal membership* functions (see Fig. 4).

As usual in fuzzy logic, every single feature can simultaneously belong to more than just one term-set (or class); the degree by which a variable belong to a particular class is expressed by the *membership value* of the related fuzzy set.

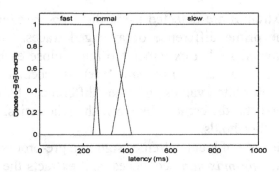

Latency ∈ {Fast, Normal, Slow}
Amplitude ∈ {Little, Normal, Big}
dRatio ∈ {Low, Normal, High}

Fig. 4. An example of *trapezoidal* fuzzy set composed by three *membership* functions; on the left side, the three wholes used in the features fuzzification

Then, the *fuzzified* values of the variables *latency* and *amplitude* are processed on the basis of a normative data table; this table subdivides the two variables in different *bands* that depend on the subject's age. For every band of age, a different Mamdani-FIS (Fuzzy Inference System) has been created.

0Acoustics P300

Band of age (years)	P300 latency (ms)	P300 amplitude (uV)
from 20 to 30	301.6 ± 26.7	9.3 ± 3.8
from 30 to 40	320.0 ± 23.4	10.1 ± 4.1
from 40 to 50	328.6 ± 24.3	9.2 ± 3.7
from 50 to 60	341.6 ± 21.3	8.3 ± 3.2
from 60 to 70	352.6 ± 22.1	7.2 ± 3.1
over 70	358.3 ± 18.9	6.9 ± 2.9

Tab. 1. Normative data related to P300 wave (acoustics stimulation)

Subsequently, a series of rules is applied to the fuzzy variables; every rules computes the likeness degree of the *min-max-min* sequence to the *ideal* P300, called *p300level* (a real number varying between 0 and 1, where a value close to 1 suggests a P300-like deflection). For an ideal P300 we assume the following characteristics: *latency*=Normal, *amplitude*=Normal, *dRatio*=Normal.

Then, the output of every rules is combined using a *defuzzification* process which allows to a global evaluation of the sequence. This process aggregates the output of all the rules by means of the *centroid* method.

Considering all detected sequences in a single trace, we classify "P300 event" the deflection that has reached the greatest likeness degree (max *p300level*).

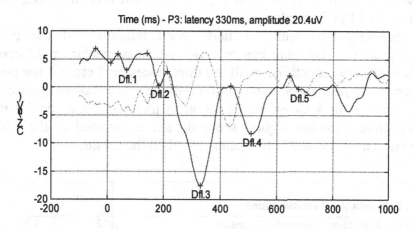

Fig. 5. The *min-max-min* sequences and the related evaluation (see Tab. 2)

0CZ deflections

0Deflection	P300 latency (ms)	P300 amplitude (uV)	dRatio	p300level
1	68	2.9	1.80	0.099
2	183	5.7	1.90	0.099
3	330	20.4	8.00	**0.793**
4	509	8.6	0.48	0.472
5	677	2.4	0.23	0.099

Tab. 2. Evaluation of the deflections in an averaged trace (Cz; max *p300level*=0.793)

Both the fuzzy sets and the implemented rules are easily modifiable and can be adapted to the interpretation of the neuro-physiologist.

5 Tests and results

The P300 recognition algorithm has been applied to 23 subjects, for a total amount of 41 trials (more than one trial is available for some subject). Every trial is characterised by the appearence of the P300 wave, and all the averaged traces for the three derivations Fz, Cz and Pz have been analysed. The considered sample is characterised by an average age of 48 years (min 18, max 86), and it is composed by 8 males and by 15 females.

Analysing the results, a fairly good reliability of the P300 identification algorithm can be observed. In particular, in 22 cases the P300 were identified in all the three traces, in 5 cases it has been recognised just in one, while in 6 cases the wave was recognised in two traces. In 8 cases the detection is incorrect, in particular the deflection with the highest score do not correspond to the P300 event. The following Tab. 3 summarised all the results.

1Total analysed examinations	**41**	
P300 detected on all traces Fz, Cz and Pz	22	53,7%
P300 detected in two traces (Fz-Cz, Fz-Pz, or Cz-Pz)	6	14,6%
P300 detected in one trace (Fz, Cz, or Pz)	5	12,2%
P300 mismatched detection	8	19,5%

Tab. 3. Results of 41 ERPs trials (three traces, Fz, Cz and Pz respectively, for every trials)

The main lack of the proposed detection support system consists into the *hard* classification, for every trace, using the greatest likeness deflection degree, even if it is very low; that is, the procedure needs a decisional element (a threshold) that allows to establish the absence of P300 wave when it is necessary.

Anyway, the developed software implements some tools for the user that can be useful for an easy analysis; more precisely, it is possible to observe in deep detail all the detected deflections, to change the analysis interval, the low-pass filter frequency, and so on.

The following arguments are put in evidence for further developments and improvements, and constitute some items for a new research:

-    the optimization of the low-pass filter cut-frequency;
-    the contemporaneous analysis of the averaged traces related to *frequent* stimuli (to get further informations useful to the P300 detection);
-    the extraction of other parameters that can be useful to the recognition;
-    the use of the EOG trace (electro-oculogram) to control the test quality;
-    the introduction of a more sophisticated decisional system to establish the absence of the P300 wave.

Moreover, we remark that most of the examined subjects suffers of neurophysiological pathologies, and do not fall within the normality field described by normative data; then, to establish more stable results for a statistical point of view, more other data will need to be examined.

6 Conclusions

The software implemented during this experimental activity, constitutes a decision support system to the physician for the P300 wave detection phase, a very useful tool for clinical diagnosis. We remark that, in this research, only the *off-line* modality was considered. Actually, this approach is quite well effective, even if some false results were obtained during the algorithm performance test. Anyway, this work suggests the basis for the *on-line* recognition of P300 wave, the most important activity not only for clinical purposes, but also for real world applications in the man-machine interface field.

164

Acknowledgements

The Authors are grateful to University of Venice for the scientific help and for the financial support to the research activity.

References

1. Abou-Chadi, F.E., Ezzat, F.A., Sif el-Din, A.A.: A fuzzy pattern recognition method to classify esophageal motility records. Ann.Biomed.Eng., 22, 1 (1994) 112-119
2. Bronzino, J.D.: Biomedical Signals: Origin and Dynamic Characteristics; Frequency-Domain Analysis. The Biomedical engineering, CRC press, cap.54 (1995)
3. Bronzino, J.D.: Digital Biomedical Signal Acquisition and Processing. The Biomedical engineering, CRC press, cap.55 (1995)
4. Berger-Vachon, C., Gallego, S., Morgonand, A., Truy, E.: Analytic importance of the coding features for the discrimination of vowels in the cochlear implant signal. Ann.Otol.Rhinol.Laryngol.Suppl.,166 (1995) 351-353
5. Molina, A., Urbaszek, A., Schaldach, M.: R-spike detection for rhythm monitoring in real time with a fuzzy logic detector. Biomed.Technol., 42 (1997) 57-58
6. Nordio, M., Giove, S., Silvoni, S.: A decision support system to prevent hypotensive episodes during dialysis. Proceedings of EMBEC'99, Graz (1999)
7. Donchin, E., Spencer, K.M., Wijesinghe, R.: The Mental Prosthesis: Assessing the Speed of a P300-Based Brain–Computer Interface. IEEE Transactions on Rehabilitation Engineering, Vol. 8, 2 (June 2000) 174-179
8. Başar-Eroglu, C., Demiralp, T., Schürmann, M., Başar, E.: Topological distribution of oddball 'P300' responses. International Journal of Psychophysiology, 39 (2001) 213-220

Evolutionary Approaches for Cluster Analysis

Sandra Paterlini, Tommaso Minerva

Università di Modena e Reggio Emilia
Dipartimento di Economia Politica
Viale Berengario 51, 41100 Modena, Italy
paterlini@unimo.it, minerva@unimo.it

Abstract. The determination of the number of groups in a dataset, their composition and the most relevant measurements to be considered in clustering the data, is a high-demanding task, especially when the a priori information on the dataset is limited. Three different genetic approaches are introduced in this paper as tools for automatic data clustering and features selection. They differ in the adopted codification of the grouping problem, not in the evolutionary operator and parameters. Two of them deals with the grouping problem in a deterministic framework. The first directly approaches the grouping problem as a combinatorial one. The second tries to determine some relevant points in the data domain to be used in clustering data as group separators. A probabilistic framework is then introduced with the third one, which starts specifying the statistical model from which data are assumed to be drawn. The evolutionary approaches are, finally, compared with respect to classical partitional clustering algorithms on simulated data and on Fisher's Iris dataset used as a benchmark.

1. Introduction

The determination of the number of groups in a dataset, their composition and the most relevant measurements to be considered in clustering the data, is a high-demanding task, especially when the a priori information on the dataset is limited. In fact, to determine the number of the g most representative groups and their composition of a dataset composed by n items with p measurements involves a time-unfeasible combinatorial effort when n and g are not small numbers and the effort is also bigger if the number of groups is not known a priori (Liu G.L., 1968).

Cluster analysis deals with data pattern detection by forming homogeneous groups inside the whole dataset by determining the groups' composition, the number of groups in the dataset, the relevant features to be used in forming the groups and a measures of similarity / dissimilarity among the items and the groups. Pre-processing data, by reducing dimensionality or detecting outliers, could be essential to remove noise, to treat with highly correlated measurements and to make the clustering algorithms more efficient to determine homogeneous groups.

To face the automatic clustering task, different algorithms have been developed, and research is still ongoing. Some of them are partitional algorithms, such as the so called *k-means*, *EM* (*Expectation Maximization*) and *fuzzy c-means* algorithms. The algorithms belonging to this class start with a random choice of the initial seeds. Then, respect to the selected seeds, compute a measure of similarity, and use this measure of similarity to determine the belonging of each observation to a specific group. Because they start from random seeds, and because of the complexity of the grouping space these algorithms not always converge to the global optimum.

To overcome this shortcoming, we propose, discuss and compare, in this paper, three clustering algorithms based on genetic algorithms (Holland 1975).

Different genetic approaches to the clustering problem have been already proposed in literature. V.V.Raghavan and K.Birchand (1979) were the first to propose to use the genetic algorithms to directly allocate the items in one of the *g* groups, which have been supposed to be present in the dataset. A fitness function directed to minimize the squared error is used to determine the optimal composition of the groups in the dataset. Since then, different genetic codification and fitness functions have been tested to solve clustering and pattern recognition problems (see Bandyopadhyay S., Murthy C.A. 1998, Srikanth R. et all 1995; Baragona, Calzini, Battaglia, 1999). Moreover, genetic algorithms have been used not only to tackle directly the clustering problem but also through the development of hybrid algorithms, that is in conjunction with other standard localized clustering techniques, such as *k-means*, *fuzzy c-means*, *artificial neural network* in order to better their performance (Tseng 2001).

The three genetic approaches we discuss in this paper, use different codification to tackle the clustering problem. We assume no a priori information is available. The grouping is determined through statistical criteria that aims at minimizing the dispersion *within* the groups and, contemporarily, maximizing the dispersion *between* the groups or maximising the likelihood of the statistical model underlying the data. The first algorithm, GAIE (*Genetic Algorithm for Items Evolution*), represents an extension (we added the capability to select automatically the number of groups and the measurements to be processed) of the algorithm proposed by Baragona, Calzini, Battaglia (1999) and Raghavan, Birchand (1979) and it is used as a benchmark to compare the erformances of the other twos. GAIE has a population of individuals which directly allocate each observation to a specific group considering the clustering problem from a combinatorial point of view.

The second, GAME *(Genetic Algorithm for Medoids Evolution),* and the third, GAPE (*Genetic Algorithm for Parameter Evolution*), are algorithms proposed by the authors. They not only exploit the genetic algorithm to determine the optimal partition, but also to get additional information about relevant grouping points in the data domain. GAME composes the groups after determining the medoids of the possible groups. For each observation GAME evaluates the euclidean distance from the group's medoids. The belonging to a specific group will be assigned to the group with minimum euclidean distance from the medoid. GAPE tackles the clustering problem in a probabilistic framework, determining the parameters of the model from which it is assumed the data are drawn.

Genetic approaches have already shown to be capable to deal with the dimensionality reduction problem and the choice of the most relevant measurements in a promising way (Raymer et al. 1997, Kim Y., Street W.N., Menczer F. 2000). We tackle this problem including into the code of the three algorithms a genetic fragment to make them able to automatic selecting the most relevant measurements to be used in clustering data.

In section 2 we briefly discuss the three algorithms and in section 3 the fitness functions we used to drive the evolutionary process. Finally, in section 4 , we discuss some empirical results and draw some conclusions.

2. Genetic Codes

Genetic algorithms are stochastic algorithm, which have been widely used in different fields because of their capability of searching the whole solution domain and of dealing with complex optimization problems. A genetic algorithm is composed by a population of individuals where each individual represents the map of a possible solution of the problem the researcher is dealing with. The population is evolved by genetic operators driven by a fitness function able to measure the degree of optimality of the individuals through the generations. The best individual coming out from the evolutionary process represents the encoding of the best approximated solution of the problem to be solved. The genetic operators (selection, crossover, mutation, elitism) are inspired to natural biological processes, driven by the Darwinian principle of the survival of the fittest individual through the generations. Genetic algorithms have numerous properties. Rudolph (1994) has shown how a genetic algorithm converges to the global optimum within an elitistic schema that is when the best individual of a generation is re-inserted in the population of the following generation. Moreover their flexibility, their parallel and straight implementation and their capability of exploring the whole search space support their validity and efficacy in numerous applications, even if they are often criticized because of their sensitivity to the control parameters (for example: the number of individuals, the number of generations, the mutation rate, the crossover rate,...).

GAIE (Genetic Algorithm for Items Evolution) algorithm directly allocates each observation of a dataset nxp in one of the g groups. Each individual string contains a fragment whose length is equal to the number of observations in the dataset and each cell can contain an integer value in the interval $[1,g]$. This codification is redundant in mapping the solution and can considerably increase the computational time required to reach the convergence resulting in a efficiency reduction of the algorithm (for example: 222111 string groups the data in the same ways 111222 string does, but they are different from the algorithmic point of view).

In the second approach (GAME) we propose to use the genetic code in order to determine some relevant points, called medoids to be used in grouping data. The *GAME* algorithm assumes that each individual is formed by $p \times g$ cells, which represents the codification of the possible values of the medoids' measurements. Each group of p cells identifies the medoid coordinates in the R^p space of the measurements. The g groups of cells represent the g medoids of the clusters. Each cell can assume a real value between the lower bound and upper bound of the whole series of the corresponding item measurements to which the medoid value of the cell is referred to. The algorithm is inspired to Forgy's approach of clustering (Forgy E.W. 1965). Once determined the possible values of the medoids of the g groups, the algorithm computes the euclidean distances between each measurement of each observation with respect to the corresponding values of the $p \times g$ medoids and determines the belonging group. Each item belongs to the cluster with minimum euclidean distance with respect to the clusters' medoids among all the computed distances between the considered item's measurements and the medoids' values of all possible groups.

Tests, comparing the speed of convergence of GAIE and GAME algorithms with respect to the same fitness function, have shown that GAME and GAIE converge to the same global optimum (if we use the same evolutionary schema and parameters), requiring a different amount of computational time. GAME converges more quickly (speed ratio around 1 to 10) and moreover can give additional information through the identification of the medoids of the dataset, which can be considered as separation points between different groups.

A probabilistic and inferential framework is introduced with the third approach. GAPE (Genetic Algorithm for Parameters Evolution) that uses the genetic codification to estimate the parameters of the statistical model which is supposed to be underlying the data and then it allocates each observation to the group with respect to which it has a higher probability of belonging. We considered the data as realization of random variables. We use the genetic algorithm to estimate the parameters in a fixed-classification model (Bock H.H. 1996). This model assumes, for a fixed number of groups g, $(G_1, G_2, .. G_g)$, a known parametric density family $f(., \theta)$ such that $X_k \sim f(., \theta_i)$ for all $k \in G_i$, $i=1,..g$ where g is unknown and the parameters vector $\Theta = (\theta_1, ..., \theta_g)$. It has been assumed that the parametric density function corresponds to the multivariate normal distribution and the genetic code is used to estimate the vector of the mean values and the diagonal elements of the variance-covariance matrix. Then, each string maps the parameters $\theta_k=(\mu_k, \Sigma_k)$.

Dealing with the full variance-covariance matrix can introduce a high number, and often statistically not relevant, of parameters to be estimated. To reduce the complexity of the problem we tried to decompose the variance-covariance matrix in a convenient way following the suggestion of Banfield and Raftery (1993). They propose to decompose the covariance matrix using an eigenvalues and eigenvector and to express it as $\Sigma_k=\lambda_k D_k A_k D_k^T$, where D_k indicates the orthogonal matrix of the eigenvectors and determine the orientation of the principal components, A_k is a diagonal matrix with elements proportional to the eigenvalues of Σ_k and determine the contours of the density functions and λ_k is a scalar that specific volume of the ellipsoids. We assume that $\Sigma_k = \lambda_k I$ and we introduce a genetic code such that three different structures of the matrix of variance are allowed. In the first case the matrix of covariance is constant among the groups and the measurements, in the second case it is constant just among the measurements and in the third case each group could have a different variance among the groups and the measurements.

The density value of each observation is computed with respect to the models determined by the genetic algorithm for each group. Each observation is attributed to the group with respect to which it has maximum density value. The fitness criteria to be used is directed to minimize the negative form of the log-likelihood of the fixed classification model:

$$F(g, \Theta) = -\sum_{i=1}^{n} \sum_{k=1}^{g} z_{ik} \log(f_{\gamma_i}(x_i | \theta_k)) \qquad (1)$$

where z_{ik} is equal to one if observation i belongs to group k or zero otherwise and where $\gamma=(\gamma_1,\ldots\gamma_n)$ are labels such that $\gamma_i=k$ if x_i belong to groups k. After processing each individual, genetic operator are applied to evolve the population to determine the best individual with associated minimum fitness value. As a result of the evolutionary process we obtained the optimum partition with respect to the adopted fitness criteria, and also the parameters of the structure underlying the data and a probabilistic measure of the belonging of each observation to a specific group.

To determine the partition of a big dataset not only requires a lot of computational effort, especially when hierarchical cluster algorithms are used, but also the analysis could be mislead by the presence of highly correlated variable and noise in the data. The genetic codification of the three algorithms includes a fragment code able to tackle this problem. The fragment is composed by p binary cells, one for each measurement. The data matrix to be processed will include just the measurements in correspondence of the cells which contain unitary values. The automatic data reduction has shown to reach a smaller number of misclassified items in the simulated dataset and in the Fisher's Iris data used as a benchmark.

3. Fitness Criteria

As previously described the GAPE algorithm uses as fitness criteria, to drive the evolutionary process, the negative form of the log-likelihood as reported in equation 2. It can be shown that in the case of constant variance matrix among the groups this is equivalent to minimize the trace of the matrix *within* the groups (Banfield J.D:, Raftery A.E., 1993).

GAME and GAIE algorithms have been tested using different fitness criteria (Calinski T., Harabasz J., 1974, Marriott F.H.C., 1982, Ricolfi L. 1992) which aim to minimize the dispersion *within* the groups and to maximize the dispersion *between* the groups. In fact the total dispersion in the dataset could be decomposed as $T=B+W$, where T indicates the *total* scatter matrix of the n observation, W the pooled-*within* groups scatter matrix and B the *between* groups scatter matrix. If we suppose apriori known the number of groups, we can use the following fitness criteria to drive the GAIE and GAME algorithms: $min(trace(W))$, $max(trace(B/W))$, $max(trace(B/T))$, $min(det(W)/det(T))$.

However, the number of groups is usually not known a priori. In this case, if we wish the algorithms could be able to determine also the best number of natural groups in the dataset, we can use two different fitness functions, which include a penalisation factor depending on the number of groups. This two fitness functions have been used to determine both the composition of the clusters and the best number of groups in the dataset. They are respectively: $min(g^2 det(W)/det(T))$ (MC, *Marriott's criterion*, 1982) and $max(\{tr(B)/(g-1)\}/\{tr(W)/(n-g)\})$ (VRC, *Variance Ratio Criterion*, Calinski T., Harabasz J., 1974). The selection of the number of groups, g, has not been included within the evolutionary process. An iterative approach, limited to the maximum available number of groups has been proposed. The algorithm selects the number of groups by finding the minimum of the fitness values obtained through the evolutionary process in which the number of groups is fixed. This mixed iterative-evolutionary schema implies the exploration of disjoint solution subspaces where the number of groups is fixed.

4. Results and Conclusions

GAIE, GAME and GAPE have been at first tested on simulated dataset composed by random observation drawn from multivariate normal distribution with different location and equivalent or not equivalent covariance structure. When data do not contain overlapping clusters, the three algorithm identify correctly the real classification. The number of misclassified items increases when the parameters used to specify the distribution from which to draw the data could lead to generate overlapping clusters. Table 1 shows the average error with respect to the real classification on simulated datasets. The average error refers to ten different simulated dataset of two hundred observations with four measurements each. Fifty observations have been generated from each multivariate normal distribution respectively with parameters: $\mu_1=[1,1,1,1]$, $\Sigma_1=I$; $\mu_2=[5,5,5,5]$, $\Sigma_2=2I$; $\mu_3=[9,9,9,9]$, $\Sigma_3=3I$; $\mu_4=[13,13,13,13]$, $\Sigma_4=4I$. For each dataset 500 simulations have been performed. Overlapping clusters may be formed. We compared our evolutionary approaches with standard techniques. GAPE, GAME and GAIE outperforms the standard *k-means* algorithm, and the *Expectation Maximization (EM)* algorithm with diagonal and full specification of the covariance matrix. Moreover, GAME, GAIE and GAPE converges to the same optimal value while the other algorithms, except the *fuzzy c-means* and the *EM* with spherical structure, tend to fall in local minima leading to variability in the number of misclassified items (see table 1). The usage of GAPE algorithm has been useful to determine the specification of the underlying model. In fact, the best individual reports values of the parameters to be estimated which are very close to the real values used to generate the data. The genetic approach not only allows to determine the optimal partition of the data, but also gives further inside about the structure of the model underlying the data.

Clustering Algorithm	Average error
GAPE	2.3%
GAIE/GAME (with MC as Fitness Function)	1.9%
GAIE/GAME (with VRC as Fitness Function)	1.5%
K-means	4.4%
Fuzzy c-means	0.9%
EM diagonal	4.6%
EM full	5.9%
EM spherical	0.7%

Table 1. Cluster algorithms performance when data are generated from multivariate normal with parameters: $\mu_1=[1,1,1,1]$, $\Sigma_1=I$; $\mu_2=[5,5,5,5]$, $\Sigma_2=2I$; $\mu_3=[9,9,9,9]$, $\Sigma_3=3I$; $\mu_4=[13,13,13,13]$, $\Sigma_4=4I$. The average error has been evaluated on 10 different simulated datasets. For each dataset 500 runs were performed so the average is on 5000 different runs.

Fisher's Iris dataset is a well-known target dataset to be used in testing the validity of new clustering algorithms. Data are collected from three different species of iris flower, where observations from just one specie have clearly distinctive features. It is composed by 150 observation with four measurements each. Table 2 report the number of misclassified items in correspondence of our evolutionary approaches compared with classical approaches and with results reported in literature. We also report results related to different fitness criteria. The results show that the evolutionary algorithms are capable to identify the correct belonging of each observation but three items, which it is the best result up to now reported. Automatic data mining, as the third column shows, allow to reduce the number of misclassified items. Moreover, respectively in the fourth and fifth columns, the best fitness values reached by the GAIE/GAME algorithm and the one in correspondence of the real correct classification are reported.

	Minimum number of misclassifed items	Average number of misclassified items on 500 runs	Misclassified Items with Automatic data-mining	Best FV	Real FV
GACE/GAIE $1/tr(B/W)$	3	3	3	7885	8930
GACE/GAIE MC	3	3	3	0.0018	0.0021
GACE/GAIE VRC	16	16	8	0.029	0.031
GACE/GAIE TrW	16	16	6	0.198	0.210
GAPE $\Sigma k=\lambda I$	16	24.5	6	1635	1693
K-means	16	25.9	---	---	---
Fuzzy c-means	16	16	---	---	---
EM-spherical	16	16	---	---	---
EM-diag	9	23.7	---	---	---
EM-full	5	17.2	---	---	---
Friedman, Rubin, 1967 $tr(W/B)$	3	---	---	---	---
Friedman, Rubin, 1967 $Det(T)/det(W)$	3	---	---	---	---
Fraley and Raftery, 1999 EM	5	---	---	---	---

Table 2. Fisher's Iris data. Comparison among the evolutionary algorithms, the standard partitional algorithms and the literature reported results in grouping the Iris dataset. The number of misclassified items with automatic data mining and not, the best fitness value reached by GAIE/GAME and fitness value in correspondence of the real classification with no automatic data mining are reported. (FV=Fitness Value; in italics are reported the Fitness Functions we used)

It should be noticed that GAME/GAIE identify a partition with associated smaller fitness values. The choice of the fitness criteria to be used is therefore crucial. A deeper look inside the data shows that the three misclassified items are more homogeneous with respect to the group they are assigned by the clustering algorithms than to the group they actually belong. The possible presence of errors in collecting data and the existence of anomalous data should be considered when looking at the partition given by the algorithms. If we compare evolutionary with classical approaches (see Table 2) we note that standard partitional clustering algorithms, which start from the random choice of the initial seeds, as GAME and GAPE do, lead at most to misclassify 5 observations and how they, except the fuzzy c-means and the EM with spherical covariance structure, can easily fall in local minima. The average number of misclassified items in 500 runs is well above the minimum number of misclassified items, showing that they are not stable in global convergence. On the contrary the evolutionary clustering algorithms show a very stable and robust convergence reaching the same state in all the 500 runs starting from different random seeds. The iterative evolutionary procedure has also determined as optimal number of groups the number of the species in the dataset in correspondence of the Variance Ratio Criterion.

The comparison between our proposed evolutionary clustering with standard clustering algorithms seems to confirm the validity of the evolutionary approach in developing more efficient clustering algorithms which have strong convergence properties as already claimed before (T.V.Le, 1995). To overcome the problem of dataset-specific clustering performance of the three algorithms we performed some tests on other datasets (wine dataset by Marais J., Versini G., van Wyk C.J., Rapp A., 1992 and Ripley's glass dataset, 1996) obtaining the same kind of results (see Paterlini, Favaro, Minerva, 2001).

These results encourage further research in improving the described algorithms and in testing and building new statistical criteria to be used. Moreover, analysis in a probabilistic framework could be further developed in testing different possible models coming from different distributions and in validating criteria, such as the BIC criterion, to automatically detect the number of groups, within an iterative approach.

References

Bandyopadhyay S., Murthy C.A., Pal S.K., "Pattern classification using genetic algorithm: Determination of H", *Pattern Recognition Letters* 19,1171-1181, 1998.

Banfield J.D:, Raftery A.E., "Model-Based Gaussian and Non-Gaussian Clustering", *Biometrics* 49, 803-821, September 1993.

Baragona R., Calzini C., Battaglia F., "Genetic algorithms and clustering: an application to Fisher's iris data", *Advances in Classification and Data Analysis* , Springer, pp.65-68, 1999.

Bock H.H., Probabilistic models in cluster analysis, *Computational Statistics & Data Analysis* 23, pp.5-28, 1996.

Calinski T., Harabasz J., "A dendrite method for cluster analysis", *Communication in Statistics,* 3(1), pp.1-27, 1974.

Forgy E.W., "Cluster Analysis of Multivariate Data: Efficiency versus Interpretability of classification", *Biometrics*, 21, 768-769, 1965.

Fraley C., Raftery A.E., "MCLUST:software for model-based cluster and discriminant analysis", *Journal of Classification*, 16, 297-306, 1999.

Friedman H.P. and Rubin J., "On some invariant criterion for grouping data", *Journal of the American Statistical Association* 63, 1159-1178, 1967.

Kim Y., Street W.N., and Menczer F. Feature selection in unsupervised learning via evolutionary search, in Proc. of the 6th ACM SIGKDD International Conference on Knowledge Discovery and Data Mining, 365-369, 2000.

Holland J.H., *Adaptation in Natural and Artificial Systems*, University of Michigan Press, Ann Harbor, 1975.

Le T.V., *Fuzzy Evolutionary Clustering*, Proceedings of the In.1 Conference on Evolutionary Computation, Perth, Nov. 29, 753-758, 1995.

Liu G.L., *Introduction to combinatorial mathematics*, McGraw Hill, 1968.

Marais J., Versini G., van Wyk C.J., Rapp A., *Effect of region on free and bound monoterpene and C13 nonrisoprenoid concentration in Weisser Riesling wines*, South African Journal of Enology and Viticulture, 13, 71-77, 1992.

Marriott F.H.C., "Optimization methods of cluster analysis", *Biometrics,* 69, 2, pp.417-422, 1982.

Paterlini S., Favaro S., Minerva T., *Genetic Approaches for Data Clustering,* Book of Short Papers, CLADAG2001, Palermo 7-8 July, 2001.

Raghavan V.V., Birchand K., " A clustering strategy based on a formalism of the reproductive process in a natural system", *in Proceedings of the Second International Conference on Information Storage and Retrieval*, 10-22, 1979.

Raymer M.L. et AAVV, "Dimensionality Reduction using Genetic Algorithms", *IEEE Transaction on Evolutionary Computation*.

Ricolfi L., HELGA *Nuovi principi di analisi dei gruppi*, FrancoAngeli s.r.l., Milano, Italy,1992.

Ripley B.D. *Pattern Recognition and Neural Networks*, Cambridge University Press, 1996.

Rudolph G., "Convergence analysis of canonical genetic algorithm", *IEEE Transactions on Neural Network*, 5(1):96-101,January 1994.

Srikanth R., George R., Warsi N., Prabhu D., Petry F.E., Buckles B.P., "A variable-length genetic algorithm for clustering and classification", *Pattern Recognition Letters* 16, 789-800, 16, 1995.

Tseng Y.L. e Yang S.B., " A genetic approach to the automatic clustering problem" , *Pattern Recognition*, Vol.34 (2), pp.415-424 (2001).

Neural Networks in Missing Data Analysis, Model Identification and Non Linear Control

S. Salini[1], .T. Minerva[2], A. Zirilli[3], A. Tiano[3], F. Pizzocchero[3]

[1] Istituto di Statistica
Università Cattolica del Sacro Cuore di Milano, Italy
(e-mail: salini.tutor@mi.unicatt.it)
[2] Dipartimento di Scienze Cognitive, Sociali e Quantitative
Università di Modena e Reggio Emilia, Italy
(e-mail: minerva@unimo.it)
[3] Dipartimento di Informatica e Sistemistica
Università di Pavia, Italy
(e-mail: antonio@controll.unipv.it)

Abstract. This paper is concerned with the problem of designing a stochastic control system for a wastewater treatment plant. Such a problem brings a sort of complexity making the task difficult to solve owing to the presence of missing data, strong interactions among the pollution variables, feedback and non-linear effects. To undertake this problem a hybrid methodology is proposed. Firstly, we addressed the problem of missing data imputation and model identification by considering classes of neural networks as potential approximated models. The optimal neural predictive model selection was obtained within an evolutionary approach (EvoNeural Model). A global search evolutionary technique is then used to establish and tune the control system related to the selected EvoNeural Model (EvoNeural Control Model). The resulting EvoNeural Control Model is then applied to a wastewater treatment plant in order to obtain an experimental validation of the proposed methodology.

1. Introduction

The most developed part of dynamical control theory deals with linear systems and powerful methods for designing controllers for such systems are currently available (Box & Jenkins, 1994). However, as applications become more complex, the processes to be controlled are increasingly characterized by strong non linear effects, interactions among subsystems, relevant stochastic components and complex information patterns. The difficulties encountered in designing control systems for such processes can include complexity, non-linearity , uncertainty.

Artificial Neural Networks (ANN, or, simply, neural networks) found wide applications in many fields, both for function approximation and pattern recognition. An Artificial Neural Network can be considered as a parametrized class of non-linear maps and numerous authors, in the last years, verified that multilayer feedforward networks and radial basis networks are capable enough to approximate a number of continuos non-linear function in a very precise sense. Narendra (1992) proposed Artificial Neural Networks as controller components in stochastic dynamical systems. In recent works Narendra and Balakrishnan (1997) and Zirilli et Al. (2000) demonstrated that ANN's could deal quite efficiently with non-linearity, complexity and noise in dynamical control systems. However the control problem can be addressed only after an efficient model identification of the system has been achieved. In a very recent work, (Salini, Minerva, Pecoraro, Pizzocchero, Tiano, 2001) the authors considered the problem of identification of multivariable dynamical model by using Artificial Neural Networks.

In this paper, firstly, we addressed the problem of missing data imputation and model identification by considering a class of Artificial Neural Networks as potential approximated models. The optimal neural predictive model was obtained within an evolutionary approach (EvoNeural Model, ENM) by establishing the input variables to be processed by the neural network as well as the network topology and properties. A global search evolutionary technique based on Genetic Algorithms is then used to evolve and tune the control system related to the selected EvoNeural Model (EvoNeural Control Model, ENCM). The resulting EvoNeural Control Model is then applied to a wastewater treatment plant in order to obtain an experimental validation of the proposed methodology.

In the next section we describe the EvoNeural Control Model, ENCM. In section 3 we show some results related to the application of the ENCM to a wastewater treatment plant and draw some conclusions.

2. Evolutionary Neural Networks for Control

The field of control is interdisciplinary in nature and extends from engineering to mathematics, from physics and chemistry to statistics involving both system modelling and statistical model validation. Today control techniques have become pervasive in a wide spectrum of applications, which are of major scientific, technological and economic importance. The objective of control is to influence the behaviour of a dynamical system to maintain the output variables at constant values (regulation) or forcing them to follow prescribed time behaviour (tracking). The control problem is to determine the control inputs to the system able to keep the output variables as close as possible to the prescribed values.

The best-developed part of control theory deals with linear systems and powerful methods for designing controllers for such systems are currently available and most of the controllers used in modern industry belong to this class. However, as applications become more complex, the processes to be controlled are increasingly characterized by the scarce availability of reliable models, multiple subsystems, high noise levels and complex information patterns. The difficulties encountered for such processes deal with complexity, non-linearity, and uncertainty so that a non-linear stochastic approach seems more adequate.

Artificial Neural Networks have recently emerged as a successful tool in the fields of sequence recognition and prediction due to the versatility in approximating an arbitrary non-linearity (Cybenko 1989). Artificial Neural Networks consist of simple calculation elements, called neurons, and weighted connections between them. Usually neurons are distributed in layers and neurons from one layer are connected to neurons in the next layer. Values are presented to the neurons in the first layer of the network (input layer), transformed by a non-linear activation function, propagated to the next layer (hidden layer) where they are weighted and biased and then propagated to the output layer. The neural weights and biases represent the parameters to be estimated in the neural model. Training neural networks means the determination of the parameters in order to obtain the best approximation of experimental outputs.

The behaviour of the net is changed by modification of the weights and bias values.

Genetic Algorithms (GAs) represent a stochastic global search and optimization tool emulating the genetic evolution of a biological population. They can be used to obtain a near optimal solution in multivariate problems without requiring the usual analytical regularity (continuity, differentiability, convexity, etc.) of the functions involved (Holland, 1975, 1995; Mitchell 1996; Goldberg 1989). The standard GA starts by choosing an initial set of candidate solutions, the population, which propagate themselves to the next generations through a set of "selection criteria". Each "individual" solution is evaluated to establish a fitness value with respect to the problem under investigation. A specific class of genetic operators (selection, mutation, crossover, cloning) are applied to the population to obtain the evolution to the next generation. The evolution ends when the population converges to the best solution matching preimposed stop criteria.

In the following we will assume that a non-linear dynamical system Σ can be described by the discrete-time state equations:

$$\begin{cases} x(t) = f(x(t-1), v(t-1); \vartheta_1) + e_1(t) \\ y(t) = g(x(t), u(t); \vartheta_2) + e_2(t) \end{cases} \tag{1}$$

where $x(t)$ represents the state vector of the dynamical system, $y(t)$ the output vector, $u(t)$ and $v(t)$ the input vectors and $e_1(t)$ and $e_2(t)$ are zero-mean gaussian white noises. We assume a complete set of discrete-time observations for the state, input and output vectors. The functions f and g and the set of parameters θ_1, θ_2 completely define a model for the dynamical system Σ. We suppose known (measured) the vector-state, x, and the input, v, at time $t-1$. The vector-state at time t, $x(t)$, will be estimated by establishing a statistical model for f. Our objective is, after establishing a statistical model for g, to determine conditions under which the input vector $u(t)$ can be found so that the outputs, $y(t)$, of the system behave in some wished fashion (the control problem).

As first step we have to consider the problem of model identification and after that we can face the problem of control system design starting from the model previously identified. If the functions f and g

are known (deduced, for example, from engineering or physical considerations about the system) the model identification implies the estimation of the parameters vector. In the case of non-linear multivariable systems, with a high number of parameters to be estimated, this can be a rather difficult task. The authors have already discussed this point in a previous work (Salini, Minerva, Pecoraro, Pizzocchero, Tiano, 2001) showing that global derivative-free optimisation techniques (such as simulated annealing or genetic algorithms) should be preferable to gradient-based algorithm. Such a minimization should, in fact, be carried out with the aim of locating the global minimum of the likelihood function, not just simply a local minimum, since only the global minimum provides an estimate with good statistical properties: unbiasedness, consistency and efficiency (Ljung,1987). In this case the control system can be obtained by restricting the class of f non linear system to these whose local linearizations are well behaved at an equilibrium state. This requires some regularity properties of the functions f and g at least in a neighbourhood of the equilibrium point. (Narendra, 1992). More interesting is the case if f and g are unknown. In such a case the model identification should be carried out only starting from the experimental observations. In this case neural networks can be considered as a class of candidate approximate models. However, the problem of how to build the best neural model is still an open issue as is the related problem of variable selection. The usual backward elimination algorithms (or forward inclusion) cannot be applied because of the path dependence induced by the non-linear relationship among the variables. To overcome the problem of model selection in a non-linear environment we used the evolutionary hybrid technique integrating Genetic Algorithms and neural networks to build an optimal (or near-optimal) model, automatically, starting from the empirical data set (Minerva, Paterlini and Poli, 1999; Salini, Minerva, Pecoraro, Tiano, Pizzocchero, 2001). In this case the evolutionary algorithm starts from a candidate set of variables and builds the model by establishing the subset (cardinality and elements) of the input variables to be included into the model. Contemporarily, the algorithm selects the network topology and the training scheme choosing from a large number of possible combinations including radial basis networks and recurrent architectures. Neural networks with feedback, in fact, can in some cases provide significant advantages over purely feed forward networks. The feedback allows to represent the state information for recursive computation, that is particularly important in the control area in which we wish to model the forward or inverse dynamics of a complex interacting system.

We call EvoNeural Model (ENM) a neural predictive model obtained within this evolutionary approach. In our case we should consider a model composed by two neural networks: one as an approximation of function f, one as an approximation of function g. If the evolutive algorithm selects neural networks N_f and N_g as an approximation for functions f and g respectively ENM of the system is:

$$\begin{cases} x(t) = N_f(x(t-1), v(t-1)) + e_1(t) \\ y(t) = N_g(x(t), u(t)) + e_2(t) \end{cases} \Rightarrow \begin{cases} \hat{x}(t) = N_f(x(t-1), v(t-1)) \\ \hat{y}(t) = N_g(\hat{x}(t), u(t)) \end{cases} \tag{2}$$

where $\hat{x}$ and $\hat{y}$ are the estimated values. Once N_f and N_g have been trained to approximate the functions f and g we can proceed exactly as they were known by a local linearization or by imposing some strong regularity conditions on N_f and N_g in order to establish the non linear feedback control law, γ, of the system: $u(t) = \gamma(x(t), y^*(t))$ where $y^*(t)$ is the reference output at time t. The theoretical existence of such a control law is a crucial step in designing the control system. If we suppose the existence of the inverse (at least in a neighbourhood of a stationary point) of f and g (or N_f and N_g), since the system is unknown, all the properties of the system (order, degree, stability, controllability, linearizability) should be deduced from the neural maps N_f and N_g. Narendra (1992, 1997) proposed to approximate the map γ by a neural network $N_\gamma(x(t), y^*(t))$. However the problem of invertibility, local linearization, strong regularity still exists. In this paper we propose an evolutionary approach to directly found a solution to the neuro-control equation:

$$y^*(t) = N_g(x(t), u(t)) = N_g(N_f(x(t-1), v(t-1)), u(t)) \tag{3}$$

The evolutionary approach uses genetic algorithm as a global search tool to tune an input vector $u(t)$ to give an output $y^*(t)$ at time t, known the vector state *x(t-1)* and the input vector *v(t-1)* at time *(t-1)*. Once the model has been identified by training the neural networks N_f and N_g on a complete set of state *(x)* , input *(u)* and output *(y)* data we set a genetic code for *u* able to represent the possible states for the input vector in the input domain. A genetic evolutive schema will be applied to search the input vector values, *u(t)*, able to maximize the fitness function represented by:

$$F\big(u(t); y*(t), x(t-1), v(t-1)\big) = -\big\| y*(t) - N_g\big(N_f\big(x(t\text{-}1), v(t\text{-}1)\big), u(t)\big)\big\| \quad (4)$$

where *x(t-1)* and *v(t-1)* are obtained by measurements on the system at time *(t-1)* and *y*(t)* is the wished output vector at time *t*. This control technique is quite similar to *'manual'* tuning where a solution is not analytically derived but is chosen by *'experimental'* selection respect to the ability to obtain the correct output. This approach *bypasses* the analytical problem of model invertibility and functions regularity by directly exploring, in a evolutive sense, the input space but does not solve the problem of the uniqueness of the solution. In fact, in a multivariable non-linear environment we can have multiple solutions to eq. 3. In our approach multiple solutions do not represent a limitation to the model invertibility since we do not use inverse function to find the input state but is still an open problem. In this paper we do not discuss deeply this point focusing on the task of finding *a* solution. The evolutive algorithm can, owing to its intrinsic parallelism, deal with multiple optimal vector states but the selection of the *'optimum inter optima'* should be carried out on criteria external to the system (for example: cost of the solution, time needed to set up, physical constrains, etc…). We'll discuss these points in a next paper. Another problem introduced by our evolutionary approach is represented by the computational time to obtain a solution. Evolutionary algorithms are known to be very slowly convergent and extremely slower respect to analytical approaches. We show in the next section that a correct genetic code can bring to a solution within few seconds on standard PCs. So we must claim that this approach cannot be applied to high frequency data but, surely, can be used to on-line control systems where the sampling interval is longer than 30 seconds. To delimit the search space we assume (as it is reasonable) that the system is BIBO-stable (BIBO, Bounded Input Bounded Output) while to speed up the convergence we use a contaminated evolutive schema where some *contaminant* individuals are introduced (*contamination*) into the initial random population. The *contaminant* individuals represent roughly approximate good solutions. They are directly selected from the past experience related to *similar* input-output-state conditions. In the next section we describe an application of the neuro-evolutionary model selection (ENM) and the neuro-evolutionary control system (ENCM) to a wastewater treatment plant. The available dataset was divided into 4 sets. Three of them were used to identify the neuro-model of the system following the usual partition into disjoint training, validation and test sets. An early-stopping training on the validation set has been used to train the neural networks. The test set has been used to measure the performance of the neural networks. This approach can take under control the problem of neural overfitting. The fourth set of data (control set) has been used to test the control system by comparing the predicted input vectors with the observed ones.

3. Application and Results

Wastewater treatment is a rather complex technological task, the efficient management of which requires the determination of reliable models. It has been recognized in the recent years that a model derived on the basis of physical laws, such as, for example, transfer kinetics, is quite difficult to be used for prediction and control purposes. In fact there is involved a quite huge number of parameters of difficult and generally not unique assessment. Furthermore, the information quantity to be processed may be quite often not available as far as monitoring routines are carried out on a real plant. In this context a statistical approach to the model selection problem seems, therefore, to be preferable since it can be directly deduced from experimental data by an identification procedure. In this paper the model identification and control problem of a large wastewater treatment plant of Northern Italy, is carried out, which is based on experimental data. The data consist of daily time series of biological concentrations, water temperature, pH, water volume per day at both input and output of the plant for a period of eight years (1984-1992). The problem can be so summarized: given the time series of biological concentrations construct a predictive and control model. This problem brings a sort of complexity that makes the task difficult to solve owing to factors such as strong interactions between the polluting variables, feedback and nonlinear effects, missing data and high dimensionality.

In our application one of the major difficulties is caused by the presence of missing values. Missing value can be due by not detected data for periodically automatic station stop, not recorded data for occasional mechanical failure or ordinary maintenance, data removed by technicians because are clearly anomalous and so on. Statisticians know that regression models, time series, neural networks and so on need complete data sets to be performed. The most popular solution to this problem belongs to the collection of imputation (fill in) techniques: last value, mean value, moving average, cubic spline, linear interpolation, EM algorithms and more again. Neural imputation, here proposed, is a technique that considers the prior knowledge about the system (in this case for example uses the information after settling to substitute the missing data in the input wastewater) and the intrinsic relationships between the variables. There are a lot of justifications for Neural Networks techniques as a method of imputation. As an example, Neural Networks create optimal homogenous groups which can be used effectively as the imputation classes, are assumption-free models, can be used as a nonparametric regression model without the assumption of specifying the true structural form of the model, provide classification imputation for a categorical dependent variable, offer potentially more reliable imputation estimates due to its ability to accommodate larger sample sizes and so on. In this approach, the method, similar to the regression-based imputation, is a sort of nonparametric regression. Missing values are replaced by the outputs predicted from a feed forward neural network. The networks are trained by using a complete-case analysis data set, using matching variables whose counterpart is present. The type of net depends on the variables in the output nodes (continuous, categorical, qualitative) and on his dependency on matching variables. The biological wastewater treatment system (see Figure 1) is composed by an input basin (waste water), a settling basin, an aeration basin and an output basin (treated water). Data on some chemical and physical variables describing the concentrations of pollutants of the water are daily collected. They are: carbon oxide, organic components, solid component in sediment and in suspension phase, nitrogen, pH and temperature. Because of the presence of very sparse data in the organic components we focused our attention only on carbon oxide and solid components and this two variables, together with pH and temperature, describe the state of the wastewater. We consider two subsystems. The first one is the *'settling'* subsystem. In this case we consider the *'after-settling'* state of the water as the state-vector of the sub-system (*x(t)*) assuming it depends on the state at a lagged time (*x(t-1)*) and on the quality of water (including pH and temperature) coming into the settling basin (*v(t-1)*). To estimate the *after-settling* state of the water at time *t* we optimize an ENM model, N_f. In this ENM model the *after-settling* state variables at time *(t-1)* (carbon oxyde and solid components) and the wastewater variables at time *(t-1)* (carbon oxyde, solid components, pH, temperature) are the input for the neural network while the state-vector, *x(t)*, is the output.

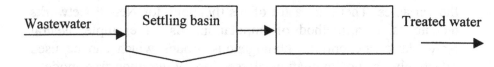

Fig. 1. The biological wastewater treatment system

The second subsystem is the *aeration-subsystem*. This is the subsystem including the control system. The output of the *aeration-subsystem* is the treated water. In the aeration basin some solvents and oxygen are introduced to speed-up the process. These variables are considered as the control variables of the system since they are determined by an external controlling system and modification of them can change the composition of treated water.

Our goal is to construct a system that indicates which and how many solvents and oxygen (treatments), *u(t)*, must be applied in order to obtain a specific level of pollutants in the water. In this case we consider the *'after-aeration'* state of the water as the output-vector of the system, *y(t)*, assuming it depends on the *after-settling* state, *x(t)*, and on the quantities of solvents and oxygen introduced into the aeration basin, *u(t)*.

$N_f \rightarrow$state vector: *x(t)*	R Cubic Spline	R Neural Imputation	$N_g \rightarrow$output vector: *y(t)*	R Cubic Spline	R Neural Imputation
Carbon oxide row	0.69	0.83			
Carbon oxide filtered	0.74	0.81	Carbon oxide row	0.73	0.87
Solid in suspension	0.61	0.77	Carbon oxide filtered	0.72	0.88
Temperature	0.92	0.98	Solid in suspension	0.84	0.79

Table 1. Results of Evo Neural Models. For each variables in the state vector and in the output vector the table shows the value of R for cubic spline imputation and neural imputation

To estimate the *after-aeration* state (content of carbon oxide and solid components in the treated water) we optimise another ENM model, N_g, where the state variables estimated by the N_f model, and the control variables (solvents and oxygen) represent the input of a neural network whose output is the *after-aeration* state, *y(t)* (output-state). The two ENM models were selected and tested by using the usual partition into disjoint datasets for neural network training, validation and test. The training and validation sets were used to estimate the network parameters and contemporarily controlling overfitting problems (early stopping applied on validation set).

The performance of the ENM was obtained by comparing the ENMs estimated output with experimental data in the test set. For both, N_f and N_g, models we obtained a very high value of R (higher than 0.8) for all variables and tested the normality of residuals. Table 1 shows the results for the neural identification models by comparing both cubic spline and neural imputation missing values techniques. The neural imputation technique allows to obtain better results in terms of model explication rate. The values of R are higher for the majority of the variables except for the solid components in suspension phase in the N_g model. A statistical check of model identification has been carried out, for each variable, by studying the regression plot between estimations and observation and by the residuals plot , the comparison between fitted and experimental series and the normality plot of residuals. In Figure 2 we show an example (for carbon oxide) of residual and fitting analysis.

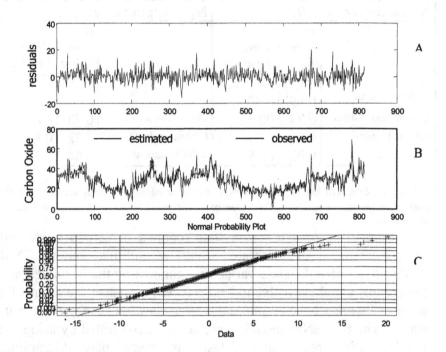

Figure 1. The graph shows the error plots (A), a plot in which predicted and observed values of carbon oxide (B) are compared and the normality plot of residuals (C).

In a previous work (Salini et al., 2000) we considered an identification model in which only one network was trained to approximate the wastewater treated system. The net had in the *input node* the input vector v, the state vector x and the control vector u and in the output node the output vector y. Table 2 shows the results in terms of R for this identification model.

Output vector : $y(t)$	R
Carbon oxide row	0.87

Carbon oxide filtered	0.88
Solid in suspension	0.79

Table 2. Results of R obtained by using only one network for the identification problems

The values of R shown in Table 2 are very close to the ones shown in Table 1 for the case of neural imputation. Using a model composed by the two nets N_f and N_g give us the advantage to estimate the state vector $x(t)$ at a future time and to compare it with the measurements. This check on the state vector is very useful to trace the wastewater treatment systems. The data used to calibrate the control system contain a complete set of measurements of the wastewater vector-state (the input state for the *settling subsystem*), the *after-settling* state (the vector state), the *control variables*, and the *after-aeration* state (the system output). This is a controlled system since the *after-aeration* state depends on the amount of solvents and oxygen introduced into the aeration basin. Here we can talk about a *simulated* control system since we cannot directly control the system but we can only learn from the data how to control it. To estimate the ENCM controller we coded the *control-variables* into a genetic algorithm chromosome. Genetic algorithms (GA) can, in fact, be used to obtain a near-optimal solution in multivariate problems without requiring the usual analytical regularity (continuity, differentiability, convexity, etc.) of the functions involved (Holland, 1975, 1995; Mitchell 1996; Goldberg 1989). The GA starts by choosing an initial set of candidate solutions, the population, which propagate themselves to the next generations through a set of "selection criteria". Each "individual" solution is evaluated to establish a fitness value with respect to the problem under investigation. A specific class of genetic operators (selection, mutation, crossover, cloning) is applied to the population to obtain the evolution to the next generation. We set up a population composed by 100 individuals and we used *contamination* (insertion of a-priori good solutions into the initial random population) as evolutionary scheme to accelerate the process. The evolution ends when the population converges to the best solution matching pre-imposed stop criteria. In our case the fitness function has been represented by the eq. 4 and the genetic evolution has the goal to find the values of $u(t)$ able to minimize the F function. In this case we know the minimum of the fitness function (zero) so as stop criterion we imposed to reach at least the approximation of 0.01 or tu run for at least 1000 generations. The 1000 generations limit was never reached. The algorithm reached the approximated convergence within 50-100 generations. The best genetic solution is the ENCM controller.

	Solvent 105	Solvent 550	O2
Mean	1720	1448	1.8
	(1743)	(1386)	(1.7)
Std	364	238	0.6
	(369)	(235)	(0.5)
Min	1000	914	0.5
	(978)	(885)	(0.1)
Max	2720	1913	2.9
	(2897)	(1916)	(3.1)

Table 3. Range Values of estimated solvents and oxygen (control variables) of the ENCM controller compared with the experimental ones (in parenthesis).

One of the problem associated to the ENCM controller is the physical compatibility of the estimated values respect to the system stability. In Table 3 we compare the mean, standard deviation and range of the estimated controller respect to the experimental one. We can note that the values of solvents and oxygen of ENCM controller are very close to the values in the data set (as we setup a *physical compatible* search space for the genetic algorithm) indicating that the solutions are in the range of likely values.

To evaluate the performance of the system we compared the estimated ENCM controller with the observed one (in this case the controller is the output of the control system) imposing that the controlled *after-aeration* state will be as close as possible to the experimentally observed state. For each control variable we evaluated the error plots, a plot in which predicted (ENCM control state) and observed values are compared, the normality plot of residuals and a regression plot between estimated ENCM controller and experimental values with the corresponding value of R (see Table 4).

Control Variable	R	Std residuals
Solvent 105	0.66	15.3
Solvent 550	0.69	11.6
O_2	0.59	0.13

Table 4. R and Standard Deviation between estimated ENCM controller and experimental values

We observed a value of R near to 0.6 for all the control variables. This not very high value of R suggests some considerations. The solution may be not unique, i.e. there may exist different combinations of solvents and oxygen rate that realize the same approximated level of polluting in the water. The difference between predicted ENCM and observed ENCM partly depends on the accuracy of the estimation of N_f and N_g. So in a more accurate data set without missing values or if we set up a more accurate identification model the results may be better. A further, and important, step will be an experimental test were allow the ENCM controller to control a real (or a scale model) wastewater plant and to compare the real output of the system with the reference output. Because of the CPU time required to obtain a control estimation, the proposed technique cannot be applied in high frequency plants. Infact, in terms of CPU time the estimate of the control vector needs on average 6 CPU minutes by using the Matlab Environment in a Windows System, Pentium III, 600Mhz, 128MRAM and about 2 CPU minutes by using Matlab in a Linux System, Pentium III Biprocessor, 800Mhz, 1024 MRAM.

The results we obtained in this simulation study encourage us to go further toward a real control experiment.

References

BOX GEORGE E. P. AND JENKINS M., (1994), *Time series analyis: forecasting and control*, Englewood Cliffs, N. J. : Prentice Hall

CYBENKO, G.,(1989), Approximation by superpositions of sigmoidal function, *Mathematical of control, Signals, and sistems*, 2, 303-314

GOLDBERG, D.E. (1989), *Genetic Algorithms in search, optimisation and machine learning*, Addison-Wesley Publishing Co

HOLLAND, J.H. (1975), *Adaptation in Natural and Artificial Systems*, University of Michigan Press, Ann Arbor

LJUNG, L. (1987). *System Identification : Theory for the User*, Prentice-Hall, London

MINERVA, T., PATERLINI, S. and POLI, I, (1999), Hybrid algorithms for time series analysis: applications to financial data, *accepted for pubblication by Economics and Complexity*, 3-4, 57-77

MITCHELL, M. (1996), *An introduction to Genetic algorithms*, MIT Press, Cambidge, MA

NARENDRA K.S. (1992), Adaptive control of dynamical systems using neural networks, *Handbook of Intelligent Control*, New York, Van Nostrand Reionhold, pp.141-183.

NARENDRA K.S. AND BALAKRISHNAN S. (1997), Adaptive control using multiple models, *IEEE Transaction on Automatic Control* (AC-42),pp.171-187.

SALINI S., MINERVA T., PECORARO A., PIZZOCCHERO F., AND TIANO A. (2001) , Comparison between neural net models and stochastic models on identification of a wastewater treatment plant, *AMSDA2001*.

ZIRILLI A., ROBERTS G.N., TIANO A., SUTTON R. (2000), Adaptive steering of a containership based on neural networks, *Int. J. Adaptive Control and Signal Processing*, (14), pp.849-873.

Genetic Optimization of Fuzzy Sliding Mode Controllers: An Experimental Study

Mariagrazia Dotoli, Paolo Lino, Bruno Maione, David Naso, Biagio Turchiano

Dipartimento di Elettrotecnica ed Elettronica, Politecnico di Bari,
Via Re David 200, 70125, Bari, Italy
dotoli@poliba.it

Abstract. Fuzzy Sliding Mode (FSM) techniques are successful in controlling nonlinear plants and reducing both control action and computational effort. However, their design is non-trivial, since it involves choosing the sliding parameter affecting the overall control speed, the input/output scaling gains, as well as the membership functions of the signal labels. This paper describes a procedure applying genetic algorithms to optimize the design of FSM controllers. Besides some simulation results, various laboratory experiments are reported, where the designed controllers are applied to the stabilization of an inverted pendulum. We employ the proposed methodology with the objective of stabilizing the pole in the upwards unstable position while simultaneously controlling the connected cart and minimizing the settling time, the cart travel and the required control action. Several conclusions are drawn out, with regard to the controller complexity and the system performance.

1 Introduction

Fuzzy Sliding Mode (FSM) control techniques are hybrid methodologies combining Sliding Mode (SM) and fuzzy control algorithms, that have recently drawn the attention of the scientific community. FSM Controllers (FSMCs) join linguistic rules and a fuzzy inference mechanism into a non linear control law, approximating the input/output map of a traditional SM controller. This results in a smooth control action, less liable to those SM noisy oscillations known as chattering, and in a system response featuring the robustness characteristics typical of SM algorithms. To design both SM and FSMCs, it is necessary to assign a linear switching surface containing the equilibrium point attracting the system trajectory. The choice of the surface slope, called the sliding parameter, is critical to the tracking convergence speed. Moreover, assigning the fuzzy controller Membership Functions (MFs) and tuning their parameters is a non-trivial task. Finally, tuning the input and output gains is as important as in fuzzy controllers.

In the literature, the design of FSMCs is traditionally based on a trial and error tuning procedure. The lack of guidelines in such a design process justifies the search for automatic design and tuning techniques. In this paper we devise two FSM control algorithms differing in complexity and performance, and subsequently optimize them via Genetic Algorithms (GAs). At present, GAs are among the most powerful tools for parametrical optimization in problems with multi-modal objective function and high computational complexity. The genetic optimization of the FSMCs devised is performed with reference to the sliding parameter, the MFs and the input/output gains; the study is carried out for both triangular and gaussian MFs.

Besides some simulation results, various laboratory experiments are reported: in the tests we employ the designed controllers for the stabilization of an inverted pendulum. This is a common benchmark task for the investigation of automatic control techniques, since the pendulum nonlinearities and instability make it a reference case study. It is to be remarked that we applied the proposed methodology with the objective of stabilizing the pole in the upwards unstable position while simultaneously controlling the connected cart and minimizing the settling time, the cart travel and the required control action.

The paper is organized as follows: section 2 reports the general concepts of SM control. Section 3 and 4 introduce respectively our FSM control technique and the case study adopted. Section 5 describes the genetic optimization of the FSMCs devised and section 6 reports several simulations and experimental tests. Finally, in section 7 some conclusions and suggestions for further research are drawn out.

2 Sliding Mode Control: A Brief Report

The SM control theory [7] was successfully employed for over a decade. Consider a single-input second-order dynamic system in the state-space form:

$$\ddot{x}(t) = f(x(t)) + g(x(t)) u(t) \qquad (1)$$

where $x = [x \quad \dot{x}]^T$ is the completely observable state vector, $f(x(t))$ and $g(x(t))$ are nonlinear functions and $u(t)$ is the control input. Furthermore, consider a given desired trajectory $x_d(t)$ and the tracking error $e(t) = x(t) - x_d(t)$ of the state component x. The basic idea of SM theory is to force the system, after a reaching phase, to a sliding surface, called sliding line for a second-order system, containing the operating point and defined as follows for system (1):

$$s(x(t)) = \left(\frac{d}{dt} + \lambda\right) e(t) = \dot{e}(t) + \lambda e(t) = 0 \qquad (2)$$

where the sliding constant λ is a strictly positive design parameter.

We must design a control action forcing the system onto the desired trajectory. Note that at steady state the system follows the desired trajectory once $s(x(t)) = 0$, i.e. when the trajectory is on the sliding line. In fact, the sliding equation (2) has a unique solution $e(t) = e(0)e^{-\lambda t}$. Since λ is positive, the steady state error is zero and the trajectory is reached. The tracking task is thus accomplished by designing a control law that forces the trajectory onto the sliding line. To this aim, we define a Lyapunov function of $s(x(t))$:

$$V = \frac{1}{2} s^2(x(t)) \qquad (3)$$

and a control law $u(t)$ such that the sliding surface is attractive, i.e. $\dot{V}$ is definite negative in the origin of the $(e, \dot{e})$ plane. This occurs if the sliding condition holds [7]:

$$\dot{V} = s(x(t)) \dot{s}(x(t)) \le -\eta \, |s(x(t))| , \qquad (4)$$

where η is a positive constant affecting the reaching phase duration [7]. It can be easily shown that (4) applies when the control law is:

$$u(t) = \hat{u}(t) - K\text{sign}\ (s(\mathbf{x}(t))\,g(\mathbf{x}(t))), \quad K > 0 \tag{5}$$

with $\hat{u}(t) = -g^{-1}(\mathbf{x}(t))\big(f(\mathbf{x}(t)) - \ddot{x}_d(t) + \lambda\dot{e}(t)\big)$ and the signum function defined as

$$\text{sign}\,(y) = \begin{cases} +1 & \text{if} \quad y > 0 \\ \;\;0 & \text{if} \quad y = 0 \\ -1 & \text{if} \quad y < 0 \end{cases} \tag{6}$$

It is extremely important to choose an appropriate value of the sliding parameter λ, because it affects the error dynamics during the sliding phase, and ultimately the convergence speed to the desired trajectory. If λ is too high the system is unstable, but if it is too small the system is sluggish: it can be tuned by hand or with automatic techniques [1]. It is to remark that controller (5) shows high frequency switching (chattering) near the sliding surface, due to the signum function. Chattering can be avoided by introducing a so-called boundary layer, i.e. replacing the signum with a saturation function [7]; this, however, can compromise the tracking accuracy.

3 A Fuzzy Sliding Mode Control Technique

A straightforward method reducing the chattering effect resulting from SM control is to fuzzify the sliding surface, i.e. combine fuzzy logic algorithms [8] with the SM control methodology [6]. The distinctive marks of the resulting FSM techniques the small magnitude of chattering and a low computational effort [2].

We propose two FSM non linear control laws, differing in the types and number of inputs and their labels and, consequently, in complexity and performance:

$$u_1 = u_{\text{fuzz}1}(e,\dot{e},\lambda) = u_{\text{fuzz}1}(s)\,, \tag{7}$$

$$u_2 = u_{\text{fuzz}2}(e,\dot{e},\lambda) = u_{\text{fuzz}2}(s_N,d)\,, \tag{8}$$

where s_N and d are respectively the normal and parallel projections of s along the sliding line [6]. The two corresponding rule bases stem from heuristic conditions in the phase plane, so that the overall control surface is akin to a SM control law with boundary layer and keeps chattering relatively small with a low computational effort (the set of rules is remarkably simple). To derive the FSMCs rule bases, let us remark that when $s(\mathbf{x}(t))$ is about zero the trajectory is close to the sliding line and the control action should be about zero. Furthermore, since the plant is of type (1), it holds:

$$s\dot{s} = s(\ddot{e} + \lambda\dot{e}) = s(\ddot{x}_d - f(\mathbf{x}) - g(\mathbf{x})u + \lambda\dot{e}) \qquad (9)$$

where the notation neglects the time dependencies. If the sign of $g(\mathbf{x})$ does not change, keeping negative for instance, in order to make (9) always negative just as in the sliding condition, a big enough positive (negative) control action is sufficient when s is negative (positive). In fact, the control action sign should change on crossing the switching line, and increase when the trajectory is far from it.

Table 1. Rule base of FSMC (7)

FSMC: u_1		
	P	N
s	Z	Z
	N	P

Table 2. Rule base of FSMC (8)

FSMC: u_2		s_N				
		N	NS	Z	PS	P
d	Z	PS	PZ	Z	NZ	NS
	P	P	PS	Z	NS	N

Employing three labels both on the input s and the output u_1 of FSMC (7), we can sum up the previous remarks in just three rules as in table 1, where N, Z, P respectively stand for negative, zero and positive. The rule base of FSMC (8) is similar, but keeps into account both the normal and parallel components of s along the sliding line. It employs five labels for s_N and two for the additional input d, resulting in ten rules as in table 2, where N, NS, NZ, Z, PZ, PS, P respectively stand for negative, negative small, negative zero, zero, positive zero, positive small and positive.

Controllers (7) and (8) can be designed choosing among several inference and defuzzification operators, as well as various types and numbers of MFs, each choice resulting in a distinct control surface that may be suitable for the problem at hand. Trying all different combinations is beyond the scope of this work, therefore we choose to employ the sup-min composition, the center of gravity defuzzification and two alternative configurations of MFs, triangular or gaussian, both uniformly partitioned for each input and output [3]. These are customary choices and result in a low computational effort. Besides, we employ a look-up-table controller, in order to reduce the algorithm run-time. A further design choice concerns the sliding parameter: it is a non-trivial task, just like in a SM controller, and critical for the tracking speed. Additional parameters are the input/output gains.

It is to remarked that using controller (7) some chattering might show, especially with a high value of the sliding parameter. In fact, even though FSMC (7) displays a smoother control surface than a traditional SM, it only employs one input, with no information on the velocity with which the trajectory approaches the sliding line. A better tracking may thus require keeping the speed into account. This can be attained for instance feeding the derivative of $s(\mathbf{x}(t))$ to the controller [1]. A compromise solution is control law (8), which does not require the computation of the derivative of s, usually a troublesome task, and keeps into account both the normal and parallel distances of the operating point from the sliding line, the latter giving insight on the distance from $e(t)=0$ and ultimately on the convergence speed.

4 Experimental Case Study: Inverted Pendulum Stabilization

In this section we describe the well-known problem of stabilizing an inverted pendulum. A pole, hinged to a cart moving on a track, is balanced upwards by a force applied to the cart via a motor. The cart is simultaneously motioned to a square wave reference position on the track, which is finite. The system state is $\mathbf{x} = \begin{bmatrix} x_1 & x_2 & x_3 & x_4 \end{bmatrix}^T = \begin{bmatrix} x & \theta & \dot{x} & \omega \end{bmatrix}^T$: it includes the cart horizontal distance from the track center, positive in the right direction, the pole angular distance from the upwards equilibrium position, positive clockwise, and their derivatives. Input u is proportional to the motor supply voltage and limited between -1 and $+1$, so that the force pushing the cart is $F=Mu$. The system model is as follows [1, 2]:

$$\dot{x} = \begin{cases} x_3 \\ x_4 \\ \dfrac{a(Mu - T_c - \mu x_4^2 \sin x_2) + l\cos x_2 (\mu g \sin x_2 - f_p x_4)}{(J + \mu l \sin^2 x_2)} \\ \dfrac{l\cos x_2 (Mu - T_c - \mu x_4^2 \sin x_2) + \mu g \sin x_2 - f_p x_4}{(J + \mu l \sin^2 x_2)} \end{cases} \qquad (10)$$

where m_c=1.12 kg and m_p=0.12 kg are respectively the cart and pole masses, $2L$=0.5 m is the pole length, g represents the gravity acceleration, J is the system moment of inertia with respect to its center of mass, f_p is the pole rotational friction coefficient and T_c is the friction affecting the cart; l, a and μ are system constants [2].

Since the pendulum model (10) is of the fourth order, the FSM technique outlined in section 3 cannot be directly applied. However, we can decouple the above equations into two second-order systems, both in companion form, representing respectively the cart and pole dynamics [2]. Thus, in order to control both cart and pole with one input, we can balance the rod first, and then the cart. In fact the pole dynamics, i.e. the second and fourth equations in (10), are affected by the cart dynamics only through friction T_c. If we neglect this contribution, we get a subsystem in the form (1). Therefore, in the following we design a FSMC regulating the pole to position e(t)=0 according to the algorithms in section 3. Note also that once the pole is balanced we get $x_2=x_4=0$ and the cart dynamics, i.e. the first and third equations in (11), are linear if we neglect friction. Linear state feedback is thus sufficient to control the cart when the pole is in equilibrium. Given the above remarks, we implement a control law via two parallel control actions [1]:

$$u=u_c+u_p. \qquad (11)$$

Control action u_c performs linear state feedback on the cart:

$$u_c = K_1 x_1 + K_3 x_3, \tag{12}$$

with gains K_1 and K_3 determined by pole allocation on the linearised inverted pendulum model [1]. The additional control action u_p in (11) employs FSM to balance the pole, i.e. control law (7) or (8). Clearly, it holds:

$$e = x_2, \quad \dot{e} = x_4, \quad s = e + \lambda \dot{e} = x_2 + \lambda x_4. \tag{13}$$

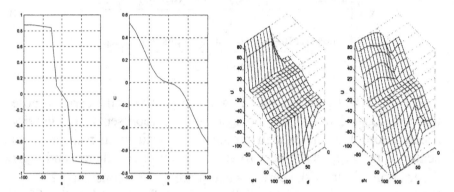

Fig. 1. Control surfaces for single-input FSMC (left, triangular; right, gaussian).

Fig. 2. Control surfaces for two-inputs FSMC (left, triangular; right, gaussian).

Note that controller (11) operates for zero initial conditions of the plant. Further, in order to implement (11) we have to assign the sliding parameter λ, the FSMC input/output scaling gains and the FSMC MFs parameters.

5 Genetic Optimization of Fuzzy Sliding Mode Controllers

Notoriously, fuzzy systems can approximate with arbitrary precision any input-output mapping by appropriately setting the membership functions, rules and inference operators. However, as in most control problems, the design task is inherently multi-criteria with conflicting objectives, and the set of parameters to adjust is too large to allow the use of heuristic knowledge or trial and error procedures, even with few rules and MFs. An emerging trend in recent research is the use of iterative search algorithms to approach such a design problem. In particular, GAs (see [4] for a throughout introduction) have been successfully employed to detect fuzzy models based on input-output data [5] and to design control laws satisfying multi-criteria performance indices [4]. GAs draw inspiration from the natural principles of the evolution of the species and the survival of the fittest. GAs work iteratively on a population of candidate solutions of the problem. The fitness, i.e. the value of the objective function associated to each solution in the population, rules the iterative selection schema, so that solutions with high fitness have high likelihood to mate and form the offspring population. Once the offspring population is selected, special operators emulating genetic crossover and mutation are randomly applied to create new solutions. These steps are then iterated in order to obtain new solutions with increased fitness until a pre-specified stopping criterion is met. Although GAs were originally designed to explore large multi-dimensional search hyper-cubes, a variety of different versions of these algorithms is available in literature to handle special cases as multi-modal and constrained problems, as the one considered in our research.

In this paper, we use a GA to optimize controller (11): in other words, the generic solution of our search problem is a FSMC. A fuzzy system is defined by the input and output membership functions, the fuzzy rules and the type of mathematical operators performing the fuzzification, inference and defuzzification processes. Although in principle it is possible to use a GA to discover and optimize both the structure (rules and operators) and the associated parameters of a fuzzy system, in order to bound the GA computational complexity and convergence time, this paper considers the optimization of the parameters of a FMSC given a fixed a priori rule base. Namely, the optimized parameters are the input/output gains, the sliding parameter and the parameters defining position and shape of the MFs (three for each). All the controller parameters are real values spanning in bounded intervals, which suggested the use of a special GA version using real valued chromosomes instead of the traditional binary encoding schema [4]. Furthermore, we developed a dynamic constraining strategy to preserve the transparency, interpretability and correctness of the optimized FSMCs. Namely, for each universe of discourse, the MFs

202

parameters can range in intervals which are dynamically computed according to the values of other neighboring parameters (e.g. the right hand vertex of the triangular MF 'Positive Medium' should be larger than the left hand vertex of the triangular MF 'Positive Big' to avoid uncovered regions of the universe). Similar constraints are added to avoid total overlap of MFs, contradictory semantics (i.e. MF 'Big' covering lower values than MF 'Medium') and other similar problems.

After some preliminary investigations on various combinations of indices of performance, we have chosen as GA fitness function (i.e. the design objective to achieve) a two-criteria index, combining with different weights the integral time absolute error (ITAE) of the cart position with respect to the set point and the integral time absolute value of the control action. Thus, the GA searches the solution space for FSMCs guaranteeing both good set point tracking and low control actions.

The final control surfaces for controllers (7) and (8) are reported in figures 1 and 2.

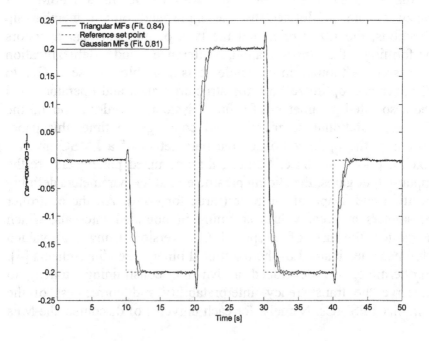

Fig. 3. Simulation results for single–input FSMC (grav. triangular: black.

6 Experimental Results

In the following we report some simulation results and laboratory tests performed on the inverted pendulum while employing the FSMCs designed in the previous sections. Note that in simulations the FSMCs were compared to a classical LQR (linear quadratic regulator) optimal controller: the FSMCs always converged faster, showing a lower overshoot.In figure 3 we report the cart position obtained in simulation employing (7) with triangular and gaussian MFs. In figure 4 the same simulation results are reported when the control law is (8). In both cases a better accuracy is remarked with gaussian MFs, especially as regards overshoot and rise time. This was expected, since the cart position fitness values estimated in Section 5 are generally lower with gaussian MFs; such a behaviour may be explained considering that both control surfaces resulting with gaussian MFs are more regular than those corresponding to triangular MFs (see figures 1 and 2).

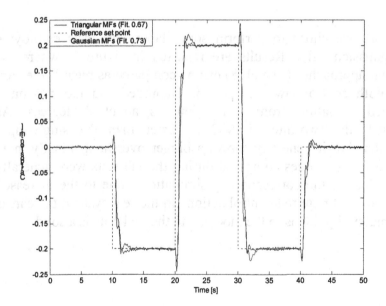

Fig. 4. Simulation results for two–input FSMC (gray, triangular; black, gaussian).

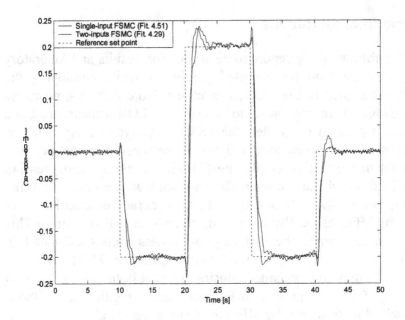

Fig. 5. Lab tests for the FSMCs (black, single-input; gray, two-input): cart position.

Therefore we chose to perform some laboratory tests employing only gaussian MFs. Results are reported in figure 5, where the different approaches (7) and (8) can be compared as regards the cart. With both control laws the pole is stabilized and the maximum observed deviation from equilibrium is about 8 degrees. As expected, the two-input FSMC is faster than the single-input controller, but it generally shows a bigger overshoot. Finally, note that the fitness values computed during the lab tests were generally bigger than in the corresponding simulations, due to the increased vibrations and required control actions in the real system: the cart is not a material point as in the model, and the real pole is a solid.

7 Conclusions

Results outlined in Section 6 show that the designed FSMCs are effective in stabilizing the pendulum and cart when the pole is initially set in its upwards unstable equilibrium.

We can drawn out several conclusions, with peculiar regard to the controllers complexity and the system performance. Both from simulations and experiments we can deduce that the two-input FSMC, even though structurally more complex than the single-input one, leads to better performance in terms of fitness and response speed. On the other hand, the single-input FSMC produces lower overshoot.

Further research directions presently under study are extending the proposed GA-optimized FSM technique to swinging-up the pole and employing a non-linear fuzzy sliding manifold both for stabilizing and swinging-up the pendulum.

References

1. Dotoli M., Maione B., Naso D., Fuzzy Sliding Mode Controllers Synthesis Through Genetic Optimization, to appear in Advances in Computational Intelligence and Learning, Methods and Applications, Zimmermann H-J., Tselentis G., van Someren M., Dounias G. eds., Kluwer Academic Publishers
2. Dotoli M., Maione B., Naso D., Turchiano B., Fuzzy Sliding Mode Control for Inverted Pendulum Swing-up with Restricted Travel, Proceedings of FUZZ-IEEE 2001 – the 10th IEEE Conference on Fuzzy Systems, Australia (2001)
3. Jantzen J., Design of Fuzzy Controllers, Rep.98E864, Danish Technical University, (1998)
4. Michalewicz Z., Genetic Algorithms + Data Structures = Evolution Programs, Springer-Verlag (1996)
5. Cordón O., Herrera F., Lozano M., A Classified Review on the Combination Fuzzy Logic-Genetic Algorithms, in Genetic Algorithms and Fuzzy Logic Systems: Soft Computing Perspectives, Sanchez E., Shibata T., Zadeh L. A. eds., World Scientific, (1997)
6. Palm R., Driankov D., Hellendoorn H., Model Based Fuzzy Control, Springer-Verlag, Berlin (1997)
7. Slotine J.-J., Weiping L., Applied Non Linear Control, Prentice Hall (1991)
8. Wang L.-X., A Course in Fuzzy Systems and Control, Prentice Hall International (1997)

Sailboat Dynamics Neural Network Identification and Control

Fabio Fossati

Dipartimento di Ingegneria Industriale e Meccanica
Facoltà di Ingegneria - Università degli Studi di Catania
Viale A. Doria, 6 – Catania, Italy
fabio.fossati@polimi.it

Abstract. A new approach to sailboat self-steering problem based on the use of neural networks is presented. Aim of the neural controller is to emulate the human steering strategies carried out by an expert helmsman. Several neural network based control strategies have been assessed: a time domain numerical simulator of a sailing yacht dynamics has been used in order to check neural controller performances.

1 Introduction

The self-steering system of a long distance cruising or racer yacht is one of the most important equipment not only for safety reasons: for boats with reduced crew, it represents not only an immediately third hand for those moments when boat crew are occupied in a sail change or manoeuvre, but without self-steering a short handed crew may find little time for sleeping, eating, and navigating. In particular for single handed ocean racing yacht, the self steering system represents the most important peace of gear she carries: in fact hand-steering is quite feasible only for a day or two; moreover for long-distance race it is a crucial device in order to reach good performance in different course, wind conditions and sea states. This situation can be reviewed as a control problem.

The most popular self-steering systems are the windvanes and the more recent electric/electronic autopilots. The first ones are totally mechanical, and they work with the aim to keep the boat with a proper direction relative to the wind, while the electronic autopilots work in order to steer the boat on a selected magnetic heading, or if they are connected to a global positioning system (GPS) on a selected course respect to the ground. Generally both systems are used on single handed ocean racing yacht because they are very different and each of them has their own advantages and/or disadvantages: the more wind blows, the better windvanes work because they have more power with higher boat speed, while electric/electronic autopilots may be used even when there is no wind or in extremely light air. Windvanes don't need electricity and more over they work even if entire electrical system is disabled; electric/electronic autopilots are cheaper and they are lighter and easier to mount.

In any case neither of these devices steers the boat like an expert sailor: in fact the "human steering" is the best system because the sailor knows how best to steer a boat on the basis of experience and intuitive opinions of the wind conditions and sea state and of boat dynamic response.

Due to these reasons, in this paper a new approach to handle the steering problem based on the use of neural networks is presented. In fact the self-learning properties of neural networks structures seem to be very attractive in order to emulate the human control strategies carried out by an helmsman.

Several neural based control strategies have been taken into account: in order to check the relevant performances a numerical simulator of a sailing yacht has been used.

In the following a summary of the simulator and neural control strategies results will be outlined.

2 Sailboat dynamics simulator

As already said in order to simulate the dynamic behavior of the sailing yacht a numerical time domain model has been used. The yacht model considers the boat as a rigid body with 4 degrees of freedom (d.o.f). Assuming a reference system as in fig. (1) the considered d.o.f. are: surge, sway, roll and yaw.

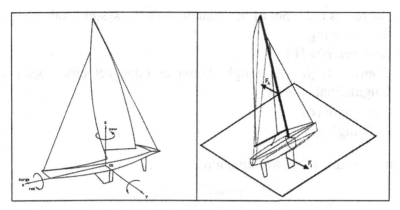

Fig. 1: Sailboat degrees of freedom and forces

In order to describe the sailboat dynamics two different reference frames have been considered (figure 2): an absolute frame with axes x_{ass} and y_{ass} in the waterplane with the y_{ass} axis in the incoming wind direction, and a local frame fixed with the boat, with the longitudinal axis x_{loc} the transversal axis y_{loc} and the vertical axis z_{loc} lying as the boat mast (fig.2). In this way the boat position is described by the center of gravity cordinates and by means of the roll and yaw angles. Forces acting on the sailboat, which have been taken into account in the present model are aerodynamic and hydrodynamic forces due to the interaction between the boat and the two surrounding fluids (wind and sea), gravity and buoyant forces and inertia forces. For a detailed description of the mathematical model refer to [1] and [3].

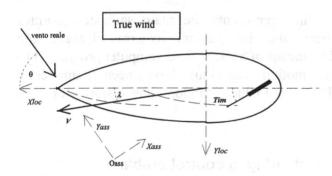

Fig. 2: Reference frames

Using this view of the "sailboat system", *wind speed and direction* can be considered as system parameters, while *rudder angle* can be considered as the input of the system; then the system state variables are the following:

- *Boat velocity* (V)
- *Leeway angle* (λ) (angle between boat velocity vector and longitudinal boat axis)
- *Yaw angle* (σ)
- *Roll angle* (ρ)

Figure 3 summarizes the situation:.

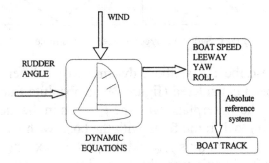

Fig. 3: Sailboat "dynamical system"

In order to take into account the steering action a regulator (PID) provides the rudder angle following simple steering strategies like cross track error, or following a specified upwind or downwind course.

Wind turbulence, which represents the main sailboat dynamics perturbation, has been taken into account in terms of space-time histories generated by means of a specialized computer program [1].

Then the sailboat motion equations have been numerically integrated, in such a way to have motion time histories of boat dynamics.

3 Optimal sailboat steering: a control problem

Generally speaking, the sailboat steering problem can be regarded as a tracking problem of a dynamical system and can be summarized as shown in figure 4:

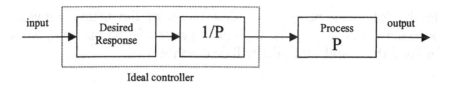

Ideal controller

Fig. 4: Ideal control process

At the generic time step the process (P) is defined by the state variables set (output) due to a certain input. Thinking to an "optimal" controller on the basis of the actual situation (wind conditions, sea state, sailing course ...) ,first of all it's able to identify the best boat performance which represents desired response and then, using the inverse process dynamics (1/P), it produces the control action which will guarantee the desired response itself.

In the human steered sailboat case, the helmsman follows just this logical scheme. In order to develop a controller able to realize this situation neural networks based control have been considered.

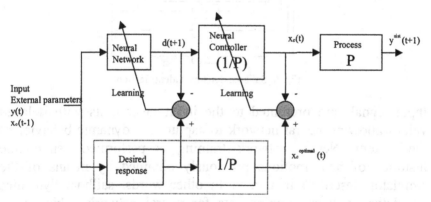

Fig. 5: Neural sailboat steering procedure: expert helmsman emulation

In principle it is possible to use a neural network able to simulate the logic block devoted to produce the desired response (reference) for the system ($d(t+1)$) and a second neural network (neural controller) which represents the inverse dynamic of the process (1/P) and able to produce the control action $x_c(t)$.

Following this approach, i.e. using two neural networks at the same time, it is possible to emulate the expert helmsman behavior (fig.5).

In the following paragraph a description of the neural networks assessed for producing the desired response will be outlined, while in Par.5 the neural controller will be described.

4 Neural network identification of sailboat dynamics

In order to assess the reference for controller at each time step, a neural network has been used able to identify the dynamic behaviour of the boat. Both multi-layer and recurrent networks have been considered. Finally a single hidden layer feed-forward neural network with 10 neurons has been successfully used.

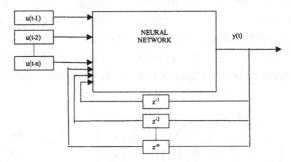

Fig. 6: Sailboat dynamics identification

Input signals are presented to the input layer units with delayed values allowing neural network to capture the dynamic behavior of the system. Neural network training is performed using time histories of boat motions previously obtained by means of the simulator described in par. 2. In other words, sailboat dynamics simulator produces training sets for neural network, which must produce the same sailboat behaviour.

A batch learning technique has been used with Back-Propagation algorithm. Input signals are boat state variables, rudder angle, wind speed and direction; output network signals are boat state variables. To guarantee convergence and for faster learning, an approach that uses adaptive learning rates has been used.

Several boat dynamics identification tests have been performed considering different sailing scenario each defined by wind speed and direction, different sailing courses and different sail sets.

As an example in the foll+owing figures boat roll and yaw angular rates time histories (fig. 7a) and boat longitudinal and lateral velocities (fig. 7b) are reported. The first 600 [s] are used as a training set for the neural network learning phase. In these figures both target data (dotted line) and identified data (solid line) are reported: in particular target data have been obtained by time domain simulator (Par.2), while identified data are produced by neural network. The two curves result perfectly superposed because learning phase has been successfully.

Considering wind turbulence a time period of 600 [s] has been considered in order to have data sets sufficiently representative of the phenomenon for network training purposes.

Once training set is completed, neural network weights are fixed and the remaining part of time histories (600-1800 [s]) represents the neural results of system dynamics identification.

As can be seen results are satisfactory.

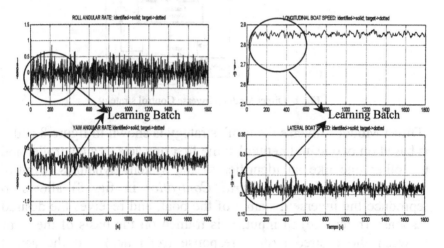

Fig. 7: (a) sailboat roll and yaw angular velocity (b) Longitudinal and lateral boat velocities

5 Neural network control of sailboat dynamics

As previously said (par. 3-4), the identification neural network trained on the basis of an expert helmsman behavior, provides the "best" desired response that represents reference for tracking control problem. Several neural control schemes have been assessed and investigated: in the following for compactness reasons only the so-called "Direct Model - Inverse Control" strategy will be described and some results will be reported. Details on the other neural control strategies used in this work can be found in [1].

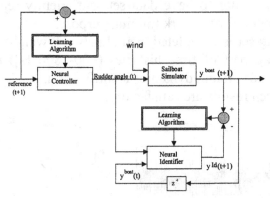

Fig. 8: Direct Model – Inverse Control strategy

"Direct Model - Inverse Control" strategy is shown in figure 8 and it is based on two neural networks: the first one acts as a controller and the other one like an identifier of the system dynamics. Control neural network (named *Neural Controller* in fig. 8) allows to reproduce the inverse dynamic of the boat, and receives the desired response (reference) in input; it is trained on the basis of the error between the desired system response (reference) and the actual system output ($y^{boat}(t+1)$ in fig. 8).

Due to Back Propagation algorithm structure, the knowledge of the system output derivatives with respect to control network weights is required: in order to evaluate these derivatives a neural network identifier has been introduced overcoming the lack of a mathematical model of the system itself. Using this neural network (named *Neural Identifier* in fig. 8) it is then possible to estimate system output (named $y^{Id}(t+1)$ in fig. 8) due to control action allowing error derivatives evaluation too.

Also in this case a single hidden layer feed-forward neural networks with 10 neurons have been used for both control and identification networks. Neural control network inputs are the desired boat response (boat state variables) and output is the rudder angle. Identification neural network inputs are boat state variable of the previous time step (t), wind speed and direction and the control signal (rudder angle) produced by neural controller, and output is the estimation of the boat actual state variables at time step (t+1).

In order to investigate performance of the proposed control strategy, the sailboat simulator described in Par. 2 has been off-line used to produce boat desired response time histories; then it has been on-line used in order produce the actual boat motion due to control action obtained at each time step by neural network controller (see fig.8). As an example in figures 9-10 a comparison between desired response (dotted lines) and actual ones (solid lines) obtained with neural controller in terms of boat speed (fig.9), apparent wind angle and boat yaw angle (fig.10) are reported.

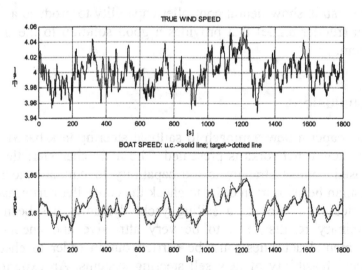

Fig. 9: comparison between reference and controlled sailboat dynamics

In fig. 9 the wind speed fluctuations time history due to turbulence effects are reported too. Note that in fig.10 in order to distinguish the target values and the actual values (i.e. values obtained with control) a time history zoom (400-800[s]) has been reported.

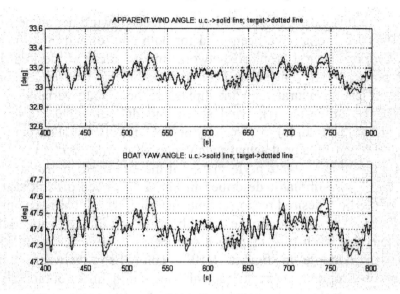

Fig. 10: comparison between reference and controlled sailboat dynamics

These results show neural controller capability to produce a control action (rudder angle) that provides a good solution to the tracking problem.

6 Conclusions

In this paper a new approach to sailboat steering task based on the use of neural networks is presented. The main characteristic of the proposed control strategy is the capability to emulate the human helmsman behaviour: in order to check the controller performances a numerical simulator of a sailing yacht dynamics has been used. Preliminary results seem to be very attractive and encouraging: experimental activities must be carried out in order to check the practical feasibility of new self-steering systems. An experimental measurement campaign with a properly instrumented ocean racer is in progress, in order to provide boat dynamics and helmsman behavior data-base for neural network controller and identifier training purposes.

References

1. Identificazione e Controllo del Comportamento Dinamico di Un'imbarcazione a Vela, *Tesi di Laurea - Dipartimento di Meccanica del Politecnico di Milano* (A. A 1999-2000)
2. R. Hecht-Nielsen: Neurocomputing, *Addison-Wesley Publishing Co., 1987*
3. F. Fossati, G. Diana: Principi di funzionamento di un'imbarcazione a vela, *Edizioni Spiegel*
4. F. Fossati, G. Moschioni: Experimental technique for determination of forces acting on a sailboat rigging, *14th Chesapeake Sailing Yacht Symposium*

Fuzzy Logic in Spacecraft Design

M.Lavagna[1], A.E.Finzi[1]

[1]Dipartimento di Ingegneria Aerospaziale-Politecnico di Milano
Via La Masa 34, 20158 Milano-Italia
lavagna@aero.polimi.it   finzi@aero.polimi.it

Abstract. This paper presents a method to automate the spacecraft preliminary design, by simulating the experts' decisional processes making use of a dedicated Fuzzy Logic Inference Engine. The problem is faced with a *multi-criteria decision-making* approach and the human expertise is captured by a variable weight vector computed by dedicated fuzzy logic control blocks implemented with a Mamdani approach.
Uncertainty of the input parameters-implicit at the starting point of a sizing process- is translated into mathematical formulation through the interval algebra rules. Comparison between simulation results and already flown space systems shew the validity of the proposed method. The results are really encouraging as the method detects space system configurations definitely similar to the real one, drastically reducing the time dedicated to the preliminary spacecraft design.
From a theoretic point of view the simulation results of the proposed method have been compared with a classic Pareto-optimal point detection to further validate them.

1 Introduction

The preliminary design of a system devoted to operate in space is a quite complex task that requires a lot of time and human effort. To answer mission requirements, coming from the payloads to be sent in space, a lot of devices must be designed - or selected among existing ones – to ensure electrical power, thermal protection, data management, in orbit insertion, instrument pointement control and telecommunication spacecraft-ground assurance. All of them highly interact and constrain the sizing process. Nowadays space systems are asked to have low mass, low power demand, low costs and high reliability.
It is then evident that, from a theoretical point of view, spacecraft design can be classified as a constrained *multi-criteria decision making* process which has to select among several alternatives the best subsystem set to answer a given goal function vector.

Currently the analysis and design of each of those subsystems is carried on by different experts' teams who work in parallel: due to the deep interaction among on-board subsystems, in fact, experts must continuously interact each others refining and correcting previous choices to achieve a final feasible solution.

At the state of the art subsystems can be modeled quite well and several software packages are available to deal with. The open problem remains the interface management among each field of analysis (e.g. electric, thermal, propulsive, informatics, structural), at each step of the project evolution, to tune each design branch on the goals, problem that is still solved by human intervention.

Human reasoning represents the decisional ring towards product optimization and it is quite difficult to be modeled as it is based on acquired expertise often made of a set of qualitative rules.

The *Concurrent Engineering* approach – focused on catching experts' knowledge by making interaction among teams and within each of them more stringent- starts being considered by the European Space Agency [1], and by Alenia Spazio with the Distributed Systems Engineering project. To automate the design process, experts systems seem to be the natural solution [2,3,4,5] and first steps start being done to apply them in spacecraft design [6,7]. An interesting approach is found in [8]: the author devotes the selection of a space propulsion system to a fuzzy decision-maker through a multi-level architecture. Starting from the last Hardy's work this paper suggests a method to automatically ranking feasible different spacecraft configurations by joining Fuzzy logic approach and Multi-Attribute Decision-Making Utility Theory (MAUT). Fuzzy Logic is devoted to capture the human knowledge and behavior in taking decisions. Human reasoning is made of logic based on proposition relationships which involve quantity qualities defined on sets with both crisp and undefined boundaries; fuzzy logic power stays in the capability of managing those qualitative relationships with a mathematical formulation – fundamental for a further output manipulation; moreover, it overcomes the Boolean representation *true-false*: different propositions at a time can be maintained true by defining their degree of satisfaction of a set of related predicates [9,10].

The method presented in the followings has the benefits of saving time and human effort giving, rapidly, a first attempt spacecraft configuration.

2 The Spacecraft Design Automation

The problem can be classified as a constrained *multi-criteria* decision-making:

$$\min \underline{G(X,R)} \quad \text{subject to} \quad \underline{C(X)} \quad (1)$$

Where:

X	=	(mx1) State variable vector	R	=	(rx1) Parameter vector
G(X,	=	(kx1)Objective/Criteria	C(X)	=	(wx1)Constraint vector

The $\underline{X}$ state variable vector elements – representing subsystem parameters - can be considered either as continuous quantities or not. In the first eventuality the problem can be managed as a Multi-Objective Decision-Making problem (MODM) to be solved with a selected multi-objective optimization method: as the solution space is unknown and the $\underline{X}$ domain is continuous the final solution might not be technologically feasible; hence, it must be compared with actually existing device parameter values for each considered spacecraft subsystem in order to find a physically feasible solution close to the optimum.

The problem becomes discrete if the $\underline{X}$ elements are limited to quantities related to existing or feasible physic devices within each subsystem field. Within a Multi-Attribute Decision-Making approach (MADM), the solution space is known hence, a voting procedure must be settled to sort the solution space elements.

Within the current problem, the last approach is better suited in terms of time saving as the final solution represents, directly, real solutions for each on-board subsystem. Hence, assuming a discrete representation for the $\underline{X}$ vector, the former (1) becomes:

$$\min \underline{G(Y(Z(X,R)))} \quad \text{subject to} \quad \underline{C(X)} \quad (2)$$

Where:

$Z|_j(X)$ = $(n(j)x1)$ alternative vector for the j-on board subsystem, $j=1,\ldots,p$

$\underline{Y}(\underline{Z})$ = (pxq) spacecraft configuration matrix from subsystem alternative combinations q, $q = \prod_{j=1}^{p} n(j)$

$\underline{G}(\underline{Y})$ = $(4xq)$ Goal matrix = [Gross Mass Required Power Cost Reliability]^T

The considered on-board subsystems are the thermal control, the power supply and storage, the propulsion, the telecommunication and the launch subsystems.

MADM methods basically create a scalar index F -called *utility function* – through a dedicated *ranking* technique. The *utility function* definition is an open problem as it is heavily *highly domain-dependent* and –as often has to collect clashing multi-disciplinary functions – can be settled through dedicated *heuristics*; classic literature shows different strategies to build that *utility function* [11]. The largely applied method, selected in this work, is the *weighted sum* strategy with s equal to one:

$$F = \left\{ \sum_{i=1}^{k} w_i^s g_i^s (x_1,\ldots, x_h, r_1,\ldots, r_v) \right\}^{\frac{1}{s}} \quad s \in N \qquad (3)$$

where: $\underline{W}$ = Weighting vector of each goal function

To evaluate the optimality of the final selected configuration, a comparison with a MODM approach as been done [11, 12].

That strategy has been chosen in order to simulate, through the *weight vector* the human way of reasoning. The $\underline{W}$ vector elements are, usually, input by the programmer or asked to the user and remain fixed throughout the process: they reflect the hierarchy the single person wants to give to each $\underline{G}$ vector element throughout the whole simulation: as a consequence, the final solution is heavily *user-dependent* and goal ordering remains stiff all over the process.

This might not be suitable for some problems that need to consider variable weights according to the related $\underline{G}$ element value. In spacecraft design, for example, an *a priori* fixed hierarchy between the gross mass and the required power criteria cannot be set as both can make a solution unfeasible; they also cannot be considered as *equal-weighted*: within each visited subsystem alternative combination both over-loaded and over-powered solutions should be highlighted: the worst goal value should drive the combination position definition in the final global ranking of the whole feasible spacecraft configurations.

For this reasons the $\underline{W}$ weight vector is here modeled to be *combination –sensitive:* $\underline{W}$ $(\underline{Y}(\underline{X}, \underline{R}))$. To that end, the $\underline{W}$ elements are computed by dedicated FLC blocks activated by each subsystem alternative present in the visited combination.

The method is going to be described in details in the following paragraphs; to better understand the fundamental role of the Fuzzy Logic within the weight element determination, an on-board subsystem analysis is going to be explained in depth.

2.1 The Global Architecture

The $\underline{G}(\underline{Y}(\underline{X}, \underline{R}))$ goal vector defined in (2) is, firstly, partitioned: the Gross Mass and the Required Power are considered completely *combination-dependent*: their mutual importance is not set arbitrarily by the user but is automatically defined by the adopted technical solutions.

Cost and Reliability, on the contrary, are *combination-free*, in the sense that their importance can correctly be considered to-be-defined by the spacecraft commissioner.

Thus a *technical* and a *non-technical* level are defined and goal and weighting vectors are split as follows:

$$\underline{G}=[\ \underline{TG}\ \ \underline{NTG}] \qquad\qquad \underline{W_i}=[\ \underline{W}_{TG,i}\ \ \underline{W}_{NTG}]\quad i=1,\ldots,q \qquad\qquad (4)$$

Where:

$\underline{TG}$ = [Gross Mass Required Power] $\qquad$ $\underline{NTG}$ = [Cost Reliability]

$\underline{W}_{TG,i}$ = $[w_{mass}\ w_{power}]_i,$ $\qquad i=1,..,q \qquad$ $\underline{W}_{NTG}$ = $[w_{cost}\ w_{rwliability}]$

The first *technical level* computes, for each new visited y_j combination, the related $\underline{TG}$ vector and the (qx2) variable $\underline{W}_{TG}(\underline{Y})$ matrix; the *non-technical level* is devoted to the $\underline{NTG}$ element definition for each y_j element, while the $\underline{W}_{NTG}$ vector is input by the user at the beginning. With the two-level approach, the equation (1) becomes:

- Level 1:

$$F_{1j} = \left\{ \sum_{i=1}^{a} w_{TG,ij} \cdot g_{ij}(x_1,...,x_k,r_1,...,r_v) \right\} \tag{5}$$

- level 2:

$$F_j = \left\{ w_{glob,NTG} \sum_{i=3}^{4} w_{NTG,i} \cdot g_{ij}(x_1,...,x_k,r_1,...,r_v) + w_{glob,TG} F_{1j} \right\} \quad j=1,...,q \tag{6}$$

The minimum F_j value specifies the best subsystem set for the current mission.

2.2 The $\underline{W}_{TG}(\underline{Y})$ Matrix

In ranking each possible combination the designers analyze each combination component, sorting each of them with respect to mass and power. To maintain that sensitivity in the automatic process, different weights are computed for each subsystem belonging to the current set: the $\underline{X}$ vector elements related to each subsystem are separately input in a FLC block which gives, as output, the $\underline{W}_j$ (kxp) matrix elements that represent the weight values of that particular subsystem within the visited combination according to each goal. Within the current application k is equal to 2 and p is equal to 4.

The $\underline{W}_{TG,j}$(kx1) vector elements, necessary for the (5), are, then, computed:

$$w_{TG,j}(i) = \sum_{s=1}^{p} W_{eig,j}(i,s) W_j(i,s) \qquad j=1,...,q \quad i=1,...,k \tag{7}$$

where:

$\underline{W}_{eig,j}$ = (kxp) matrix whose columns correspond to the eigenvectors of the maximum eigenvalue of the $\underline{B}_i$ matrix

$\underline{B}_i$ = matrix of the pairwise comparisons between the same row elements of the $\underline{Y}$ matrix with respect to the g_i goal

The $\underline{W}_{eig,i}$ columns are the scale rating of the $\underline{Y}$ elements- within each column - as to their importance with respect to each i-th *criterion*.

Defining a pairwise relevance between subsystems represented by the z components of each $\underline{y}_j$ element with respect to each *criterion* is certainly easier than directly writing an absolute hierarchy among them and filling the $\underline{\underline{W}}_{eig,j}$ matrix: thus, if b_{hj} is the comparison value between element z_h and element z_j within the g_i goal for the y_j combination, a square matrix $\underline{\underline{B}}_i$ (pxp) can be obtained. The eigenvector corresponding to the maximum eigenvalue of the $\underline{\underline{B}}_i$ matrix is a cardinal ratio scale of the p compared elements [15].

2.3 The $\underline{G}(\underline{Y})$ Goal Vector

As the $\underline{G}$ vector elements can differ both in nature and in magnitude, they are normalized through a single-membership dedicated fuzzy block. The output of this process is an index L_{ij} with values between 0 and 1, corresponding, respectively, to the worst and to the best g_{ij} value; related memberships have a trapezoidal shape and are tuned on the g_i values in agreement with the constraints on it. With rules exposed in the former and in the current paragraphs, the (5) and (6) become:

• level 1 and 2:

$$L_{1j} = \left\{ \sum_{i=1}^{2} w_{TG,ij} L_{ij} \right\} \quad L_{glob\,j} = \left\{ w_{glob,NTG} \sum_{i=3}^{4} w_{NTG,i} L_{ij} + w_{glob,TG} L_{1j} \right\} \quad j=1,\ldots,q \quad (8)$$

In order to deal with the intrinsic uncertainty the $\underline{X}$ elements always present in a preliminary design, all inputs are defined within two ranges, a *large likely value* interval (L.L.V) and a *most likely value* interval (M.L.V.) [8]: all the computational load is managed with the interval algebra rules [13]. The final $\underline{L}_{glob}(\underline{Y})$ elements are no more crisp hence a complete ordering of the feasible spacecraft configurations is unavailable. A final result ranking method has been then defined to select the suggested $\underline{Y}$ element to answer the mission requirements with the best $\underline{G}$ vector [12,14].

3 The Fuzzy Logic Role: The Power Subsystem Example

Fuzzy logic plays a fundamental role in the $\underline{\underline{W}}$ element definition and it is the core of the human thinking simulation within the proposed method. As an example the power supply and storage subsystem management is going to be analyzed.

Spacecraft power supply can be accomplished by three types of power sources, the photovoltaic solar cells, the static heat sources for direct thermal-to-electric conversion, the dynamic heat sources which make the same former conversion through a Brayton, Stirling or Rankine cycles [16]. The first type needs solar power to work and, then, it must be coupled with a storage power subsystem to maintain electric power availability during eclipse periods; the typical devices for power supply in eclipse are batteries. The power supply and storage subsystems intervene only in the gross mass goal computation, thus the related $\underline{w}_{TG,j}$ determination is reduced to the $w_{mass,j}$ definition. Hence, the designers firstly have to define whether a photovoltaic or a static/dynamic power source better fits the current spacecraft requirements. A first rough taxonomy of possible alternatives is done taking into account the mission parameters such as the average distance from the sun and the global eclipse period: the higher the first is, the larger a solar panel has to be, that means an increasing mass; the longer the second is the higher the battery number and the larger the solar panels: as batteries supply power during shadows are directly fed by the solar panels, eclipse definitely influences the whole system mass. Thus a typical driving criterion is: *" IF the sun distance is high and the eclipse period is long THEN a photovoltaic power subsystem is heavy";* hence, all the photovoltaic solutions will be considered worse than the static dynamic ones with respect to the mass.

Moreover, within each aforementioned class, sub-classes exist depending on the technological available devices: solar panels – for example - can be made either of silicon or gallium-arsenide cells. Hence, within the solar panel solution, the engineers decide to award a particular cell material depending on its efficiency conversion together with the material density as the first parameter drives the panel dimension while the second one defines its mass. Again, a typical qualitative rule is:*" IF the cell efficiency is high and the density is low THEN a photovoltaic power subsystem is light".*

Several quality classes can be defined both for parameters and outputs and several conjunctive and disjunctive premises can be generated to activated correspondent consequences. It appears, then, that an "automatic spacecraft designer" has to manage quantitative quantities -coming from a rough dimensioning process –qualitatively related to the criteria set. For those reasons the Fuzzy Logic approach has been selected as the most fitting method to faithfully reproduce the real selection process.

Coming from the former inference rules the selected $\underline{X}$ quantities to input in the dedicated MISO FLC loop, to obtain the correspondent $W_{power\ source\_mass,j}$ weight, are:

η	=	Efficiency conversion between solar and electric power (SA)
$\rho\ (Kgm^{-2})$	=	Cell density (SA)
RP (W)	=	Required power in the visited y_j combination (SA,BA,AS)
SE $(Whrkg^{-1})$	=	Specific energy (BA)
SP (WKg^{-1})	=	Specific power supplied (SA)
Ep	=	Ratio between eclipse and orbital period (AS)
d (AU)	=	Distance from the sun (AS)

The $W_{power\ source\_mass,j}$ takes place in the $\underline{W}$ matrix and - through equation (7)- takes part in computing the final $L_{glob,j}$ fuzzy index. The membership shape, number and width have been tuned on the related quantities starting from available data sheets of the devices they are related to.

The same reasoning drives each subsystem alternative judgement with respect to the other criteria and for each to be designed. Table1 presents the selected methods for each operation involved in a classic fuzzy loop. Some of them are simply a first attempt solution to check whether the global architecture works.

Operation	AND	OR	Inference	Aggregation	Defuzzification
Method	min	max	min	sum	Centroid

Tab. 1. Method selection

4 Applications

As already said two different validations have been made. The first one, applied to a simplified architecture, compared the results obtained by a classic Pareto line detection to those coming from the simulations [7]. The second one took already flying/flown space missions and compared the final subsystem set selection to that suggested by the code. Both of them gave encouraging results, as solutions are definitely similar. Table 2 shows results coming from the comparison with the NASA's Mars Global Surveyor mission. The launcher selection is not present as it is under refinement. The best combination and the near-best combination are shown: goal function values are in the same range and the differences come from simplified model for the considered subsystems and from a possible difference in considered alternatives for the on-board devices.

	MGS	Best Configuration			
		L.L.V.	M.L.V.	$W_{TG,m}$	0.60
Power	980W	960-1340W	1100-1200W	$W_{TG,p}$	0.32
Wet Mass	1062.1Kg	912.98-1416.48Kg	1008.22-1186.02Kg	$W_{glob,TG}$	0.75
Propellant Mass	388.3Kg	508.8-844.31Kg	569.32-687.86Kg	$W_{glob,NTG}$	0.25
Cost	239 M$	130-450M$	190-360M$		
Reliability	//	0,85-0,94	0,88-0,91		

Table 2. Goal function comparison with the MGS and automatically detected TG weights

Conclusions

The paper presents a method to automate the spacecraft design by simulating the humans' behavior in taking decisions. The problem is faced with *a multi-criteria decision making* approach and the final on-board subsystem set is selected by computing a preference index coming from a weighted sum of the given goals. The core of the method is the weight vector that is maintained *configuration-sensitive* during the process: its elements are computed through a nested architecture, made of several FLC blocks. Fuzzy Logic translates the qualitative rules coming from designers' expertise in a mathematical formulation without loosing the richness of the human thinking. Comparison with classic multi-criteria optimization methods as well as with already flying missions highlighted the validity of the proposed method that must anyway be refined and improved. In particular, a training process with Neural Networks is going to be considered to define rules and memberships involved in each FL loop.

References

1. M.Bandecchi, Melton, F.Ongaro: Concurrent Engineering Applied to Space Mission Assessment and Design.In: ESA Bulletin 99,1999
2. C.M.Fonseca, P.J.Fleming: An Overview of Evolutionary Algorithms in Multi-Objective Optimization. In: Evolutionary Computation, (Spring 1995) 3(1):1-16
3. R.R.Yager: Multiple Objective Decision-Making Using Fuzzy Sets. In: International Journal of Man-Machine Studies, (1977),nr.9, 375-382
4. J.Lee, J.Kuo, W.Huang: Fuzzy Decision-Making through Relationships Analysis between Criteria. IEEE, 0-7803-3687-9/96
5. U.Kaymak H.R.Yan Nauta Lemke: A Parametric Generalized Goal Function for Fuzzy Decision Making with Unequally Weighted Objectives. 0-7803-0614-7/1993 IEEE
6. A.Fukunaga, S.Chien,D.Mutz: Automating the Process of Optimization in Spacecraft
   Design. (1996) Jet propulsion Laboratory, California Institute of Technology
7. T.Mosher: Spacecraft Design Using a Genetic Algorithm Optimization Approach. 0-7803-4311-5/1998, IEEE
8. T.L.Hardy: Fuzzy Logic Approaches to Multi-Objective Decision-Making in Aerospace Applications. In: Proceedings of 30th AIAA/ASME/SAE/ASEE joint Propulsion Conference, Indianapolis-IN (June 27-29, 1994)
9. L.A.Zadeh: Fuzzy Sets. Information and Control (1965) 8, 338-353
10. G.J.Klir, B.Yuan: Fuzzy Sets and Fuzzy Logic. Prentice Hall PTR,New Jersey (1995)
11. H.Eschenauer J.Koski A. Osyczka: Multi-criteria Design Optimization. Springer-Verlag (1990)
12. M.Lavagna, A.E.Finzi: Multi-criteria Design Optimization of a Space Syestem with a Fuzzy Logic Approach. In: Proceedings of the 42nd AIAA/ASME/ASCE/AHS/ASC Conference, Seattle-WA (16-19/04/2001)
13. R.B.Kearfott: Interval Computation: Introduction, Uses,and Resources. http://cs.utep.edu/interval-comp/main.html
14. M.Lavagna,A.E.Finzi: Multi-Criteria Optimization for Space Mission Design: a New Approach. In: Proceedings of the i-Sairas Conference Montreal-CA (18-21/06/2001)
15. T.L.Saaty: A Scaling Method for Priorities in Hierarchical Structures. In: Journal of Mathematical Psychology (1977) 15(3), 234-281
16. R.X.Meyer: Elements of Space Technologies. Dept. Of Mechanical and Aerospace Engineering, University of California, Los Angeles(1999)

Printing: Strauss GmbH, Mörlenbach
Binding: Schäffer, Grünstadt